Student's Solutions Manual

Paul Lorczak

Solutions to Computer Exercises provided by
Atanas Rountev and Matthew Arnold

to accompany

Discrete Mathematics

Sherwood Washburn
Seton Hall University

Thomas Marlowe
Seton Hall University

Charles T. Ryan
Formerly of Seton Hall University

An imprint of Addison Wesley Longman, Inc.

Reading, Massachusetts • Menlo Park, California • New York • Harlow, England
Don Mills, Ontario • Sydney • Mexico City • Madrid • Amsterdam

Reproduced by Addison-Wesley from camera-ready copy supplied by the author.

Copyright © 2000 Addison Wesley Longman.

All rights reserved. No part of this publication may be reproduced, stored in a retrieval system, or transmitted, in any form or by any means, electronic, mechanical, photocopying, recording, or otherwise, without the prior written permission of the publisher. Printed in the United States of America.

ISBN   0-201-61925-3

1 2 3 4 5 6 7 8 9 10 VG 0403020100

Contents

Chapter 1	Sets, Subsets, Induction and Recursion	**1**
Chapter 2	Integers, Remainders, and the Golden Ratio	**22**
Chapter 3	Functions, Relations, and Counting	**31**
Chapter 4	Graphs	**41**
Chapter 5	Proof Techniques and Logic	**56**
Chapter 6	Boolean Algebras, Boolean Functions, and Logic	**64**
Chapter 7	Graphs and Relations	**78**
Chapter 8	Algorithms	**84**
Chapter 9	Combinatorics	**91**
Chapter 10	Models of Computation	**102**
Solutions to Computer Exercises	Chapters 1-10	**106**

Chapter 1 Sets, Subsets, Induction and Recursion

Section 1.1 Exercises

The exercises in this section reference the rows of the Pascal triangle shown here. The rows are numbered from the top starting with row 0, row 1 and so on. Similarly the entries in each row are numbered 0, 1, 2...

```
                                    1
                                 1     1
                              1     2     1
                           1     3     3     1
                        1     4     6     4     1
                     1     5    10    10     5     1
                  1     6    15    20    15     6     1
               1     7    21    35    35    21     7     1
            1     8    28    56    70    56    28     8     1
         1     9    36    84   126   126    84    36     9     1
      1    10    45   120   210   252   210   120    45    10     1
   1    11    55   165   330   462   462   330   165    55    11     1
 1    12    66   220   495   792   924   792   495   220    66    12     1
1   13    78   286   715  1287  1716  1716  1287   715   286    78    13    1
1  14   91   364  1001  2002  3003  3432  3003  2002  1001  364    91    14    1
```

In exercises 1-7 the desired binomial coefficient $\binom{n}{k}$ is computed as the kth element in the nth row of the Pascal triangle.

1. $\binom{0}{0} = 1$ 3. $\binom{5}{3} = 10$ 5. $\binom{10}{5} = 252$ 7. $\binom{12}{6} = 924$

9. $\binom{0}{0} = \dfrac{0!}{0!0!} = 1$ 11. $\binom{5}{3} = \dfrac{5!}{3!2!} = \dfrac{5 \cdot 4}{2 \cdot 1} = 10$

13. $\binom{10}{5} = \dfrac{10!}{5!5!} = \dfrac{10 \cdot 9 \cdot 8 \cdot 7 \cdot 6}{5 \cdot 4 \cdot 3 \cdot 2 \cdot 1} = 252$ 15. $\binom{12}{6} = \dfrac{12!}{6!6!} = \dfrac{12 \cdot 11 \cdot 10 \cdot 9 \cdot 8 \cdot 7}{6 \cdot 5 \cdot 4 \cdot 3 \cdot 2 \cdot 1} = 924$

17. Each of the full binary trees with three levels below illustrates one of the three paths having exactly one edge to the left. The path is indicated with heavier edges.

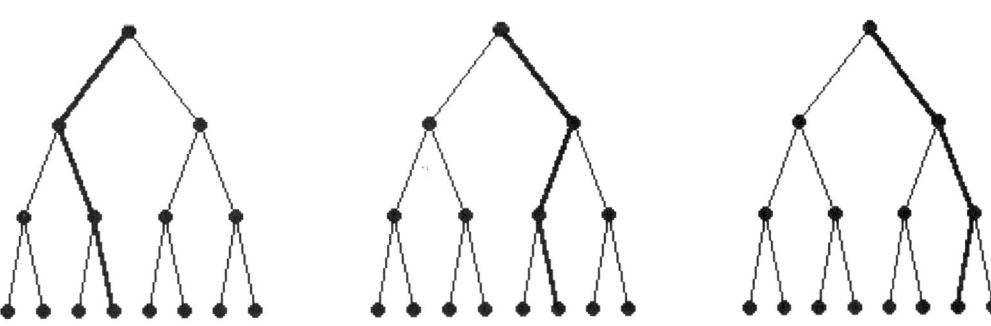

19. Each of the full binary trees with four levels below illustrates one of the four paths having exactly one edge to the left. The path is indicated with heavier edges.

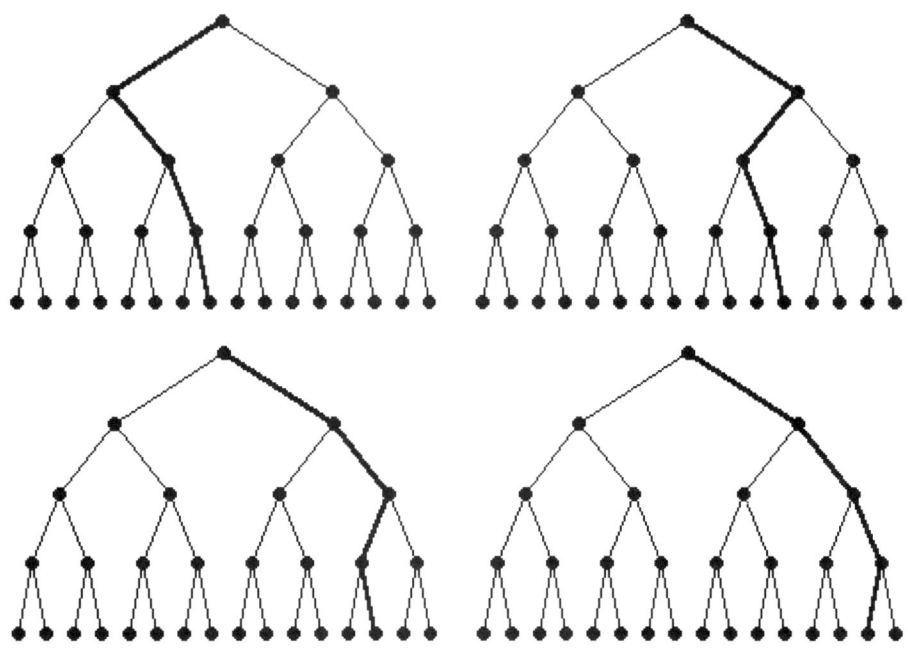

21.
$n=0\ :\ 1=2^0$

$n=1\ :\ 1+1=2=2^1$

$n=2\ :\ 1+2+1=4=2^2$

$n=3\ :\ 1+3+3+1=8=2^3$

$n=4\ :\ 1+4+6+4+1=16=2^4$

$n=5\ :\ 1+5+10+10+5+1=32=2^5$

$n=6\ :\ 1+6+15+20+15+6+1=64=2^6$

$n=7\ :\ 1+7+21+35+35+21+7+1=128=2^7$

Section 1.1 Advanced Exercises

1.

```
                                             1
                                          1     1
                                       1    2     1
                                    1    3     3     1
                                 1    4     6     4     1
                              1    5    10    10     5     1
                           1    6    15    20    15     6     1
                        1    7    21    35    35    21     7     1
                     1    8    28    56    70    56    28     8     1
                  1    9    36    84   126   126    84    36     9     1
               1   10    45   120   210   252   210   120    45    10     1
            1   11    55   165   330   462   462   330   165    55    11     1
         1   12    66   220   495   792   924   792   495   220    66    12     1
      1   13    78   286   715  1287  1716  1716  1287   715   286    78    13     1
   1   14    91   364  1001  2002  3003  3432  3003  2002  1001   364    91    14     1
```

3.

```
                                             1
                                          1     1
                                       1    2     1
                                    1    3     3     1
                                 1    4     6     4     1
                              1    5    10    10     5     1
                           1    6    15    20    15     6     1
                        1    7    21    35    35    21     7     1
                     1    8    28    56    70    56    28     8     1
                  1    9    36    84   126   126    84    36     9     1
               1   10    45   120   210   252   210   120    45    10     1
            1   11    55   165   330   462   462   330   165    55    11     1
         1   12    66   220   495   792   924   792   495   220    66    12     1
      1   13    78   286   715  1287  1716  1716  1287   715   286    78    13     1
   1   14    91   364  1001  2002  3003  3432  3003  2002  1001   364    91    14     1
```

5.

```
                                             1
                                          1     1
                                       1    2     1
                                    1    3     3     1
                                 1    4     6     4     1
                              1    5    10    10     5     1
                           1    6    15    20    15     6     1
                        1    7    21    35    35    21     7     1
                     1    8    28    56    70    56    28     8     1
                  1    9    36    84   126   126    84    36     9     1
               1   10    45   120   210   252   210   120    45    10     1
            1   11    55   165   330   462   462   330   165    55    11     1
         1   12    66   220   495   792   924   792   495   220    66    12     1
      1   13    78   286   715  1287  1716  1716  1287   715   286    78    13     1
   1   14    91   364  1001  2002  3003  3432  3003  2002  1001   364    91    14     1
```

4   Chapter 1 Sets, Subsets, Induction and Recursion

7. The entries divisible by a particular integer n form patterns in the shape of inverted triangles. This is not surprising since whenever two consecutive entries in a row are divisible by n, the entry between them in the next row, being their sum, must also be divisible by n. Further, if an entry divisible by n is followed in its row by an entry not divisible by n then their sum in the next row will not be divisible by n. These two facts lead to a triangular shape. Note also that the entries in row n will all be divisible by n with the exception of the endpoints of 1 since these entries are equal to $\binom{n}{k} = n \cdot \left(\frac{(n-1)!}{k!(n-k)!} \right)$.

Section 1.2 Exercises

In each exercise, the symbol P(n) denotes the proposition involving n to be proved by induction in that exercise.

1. Initial step: $2 \cdot 1 = 2 = 1(1+1)$ establishing P(1).
   Induction step: $2 + 4 + 6 + \ldots + 2n + 2(n+1) = n(n+1) + 2(n+1)$ (assuming P(n))

   $= (n+1)(n+2)$ proving P(n+1).

3. Initial step: $1 = 4 \cdot 1 - 3$ establishing P(1)
   Induction step: $1 + 5 + \ldots (4n-3) + (4(n+1) - 3) = n(2n-1) + (4n+1)$ (assuming P(n))
   $$= 2n^2 + 3n + 1$$
   $$= (n+1)(2n+1) \quad \text{proving P}(n+1).$$

5. Initial step: $\dfrac{1}{1 \cdot 3} = \dfrac{1}{3} = \dfrac{1}{2 \cdot 1 + 1}$ establishing P(1).

   Induction step: $\dfrac{1}{1 \cdot 3} + \dfrac{1}{3 \cdot 5} + \ldots + \dfrac{1}{(2n-1)(2n+1)} + \dfrac{1}{(2(n+1)-1)(2(n+1)+1)}$

   $$= \frac{n}{2n+1} + \frac{1}{(2n+1)(2n+3)} \quad \text{(assuming P}(n)\text{)}$$

   $$= \frac{2n^2 + 3n + 1}{(2n+1)(2n+3)}$$

   $$= \frac{(2n+1)(n+1)}{(2n+1)(2n+3)}$$

   $$= \frac{n+1}{2n+3} \quad \text{proving P}(n+1).$$

7. Initial step: $\displaystyle\sum_{k=0}^{0} 3^k = 3^0 = 1 = \dfrac{3^{0+1} - 1}{2}$ establishing P(0).

   Induction step: $\displaystyle\sum_{k=0}^{n+1} 3^k = \sum_{k=0}^{n} 3^k + 3^{n+1} = \dfrac{3^{n+1} - 1}{2} + 3^{n+1}$ (assuming P(n))

   $$= \frac{3^{n+1} - 1 + 2(3^{n+1})}{2} = \frac{3^{n+2} - 1}{2}$$

   proving P(n+1).

9. Initial step: $4! = 24 \geq 16 = 2^4$ establishing P(4))

   Induction step: Assume $n > 4$. $(n+1)! = (n+1)n! \geq (n+1)2^n$ (assuming P(n))
   $$\geq 2(2^n) \quad (\text{since } n+1 \geq 2)$$
   $$= 2^{n+1} \quad \text{proving P}(n+1).$$

11. When $t = 5$, Hypsicles' formula for the nth hexagonal number becomes $\dfrac{n(5n-3)}{2}$. Substituting $n = 1, 2, \ldots 12$ yields the sequence 1, 7, 18, 34, 55, 81, 112, 148, 189, 235, 286, 342.

13. Initial step: $\displaystyle\sum_{k=1}^{1}(kt-(t-1)) = t-(t-1) = 1 = \dfrac{1(1\cdot t - (t-2))}{2}$ establishing P(1).

    Induction step: $\displaystyle\sum_{k=1}^{n+1}(kt-(t-1)) = \sum_{k=1}^{n}(kt-(t-1)) + (n+1)t - (t-1)$
    $$= \dfrac{n(nt-(t-2))}{2} + (n+1)t - (t-1) \quad (\text{assuming P}(n))$$
    $$= \dfrac{(n+1)^2 t - (n+1)(t-2)}{2} = \dfrac{(n+1)((n+1)t - (t-2))}{2} \quad \text{proving P}(n+1).$$

15. Initial step $\displaystyle\sum_{k=0}^{0}\dfrac{k(3k-1)}{2} = \dfrac{0(-1)}{2} = 0 = \dfrac{0^2(0+1)}{2}$ establishing P(0).

    Induction step: $\displaystyle\sum_{k=0}^{n+1}\dfrac{k(3k-1)}{2} = \sum_{k=0}^{n}\dfrac{k(3k-1)}{2} + \dfrac{(n+1)(3(n+1)-1)}{2}$
    $$= \dfrac{n^2(n+1)}{2} + \dfrac{(n+1)(3n+2)}{2} \quad \text{assuming P}(n))$$
    $$= \dfrac{(n+1)(n^2+3n+2)}{2} = \dfrac{(n+1)^2(n+2)}{2} \quad \text{proving P}(n+1).$$

Section 1.2 Advanced Exercises

1. 9 moves

3. 13 moves

Section 1.3 Exercises

1. True. 1 is an element of the set {1,2,3}.

3. False. The set consisting of the single element 1 is not an element of {1,2,3}.

5. True. Every element of {1} is an element of {1,2,3}.

7. The statement $A \in A$ has no meaning. In order to have a consistent theory the question of a set being an element of itself is disallowed.

9. The statement $A \in B$ cannot be assigned a truth value without additional knowledge of A and B. For example, if $B = \{1,2,\{3\}\}$ then $A = \{3\}$ makes the statement true while $A = \{2\}$ makes the statement false

11. $A = \{a \in N: a = 2b \text{ for some } b \in N \ \& \ a \leq 12\}$

13. $A = \{a \in N: a = 4b \text{ for some } b \in N \ \& \ a \leq 24\}$

15. $A = \{a \in N: a = 2^b - 1 \text{ for some } b \in N \ \& \ b \leq 6\}$

17. $A = \{a \in N: \dfrac{1}{a} = 0\}$

19. $A = \{a \in N: a = 4b + 1 \text{ for some } b \in N \cup \{0\}\}$

21. $A \times B = \{(1,1),(1,2),(1,8),(1,9),(3,1),(3,2),(3,8),(3,9),(5,1),(5,2),(5,8),(5,9),$
    $(7,1),(7,2),(7,8),(7,9),(9,1),(9,2),(9,8),(9,9)\}$
    $A \cup B = \{1,2,3,5,7,8,9\}$
    $A \cap B = \{1,9\}$
    $A - B = \{3,5,7\}$
    $A^c = \{2,4,6,8\}$
    $B^c = \{3,4,5,6,7\}$

23. $A \times B = \{(1,4),(1,8),(1,12),(1,16),(4,4),(4,8),(4,12),(4,16),(7,4),(7,8),(7,12),(7,16),$
    $(10,4),(10,8),(10,12),(10,16),(13,4),(13,8),(13,12),(13,16),(16,4),(16,8),(16,12),(16,16)\}$
    $A \cup B = \{1,4,7,8,10,12,13,16\}$
    $A \cap B = \varnothing$
    $A - B = \{1,7,10,13\}$
    $A^c = \{2,3,5,6,8,9,11,12,14,15\}$
    $B^c = \{1,2,3,5,6,7,9,10,11,13,14,15\}$

25. $\chi(B) = (1, 0, 1, 0, 1, 0)$

27. $\chi(B) = (0, 0, 1, 1, 1, 0, 0)$

29. The path representing $\{1,3,4\}$ is the unique path beginning at the root vertex and ending with the vertex labeled $\{1,3,4\}$.

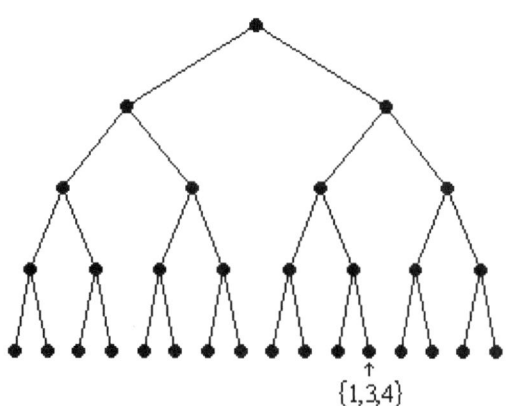

31. The path representing (1, 1, 0, 1) is the unique path beginning at the root vertex and ending with the vertex labeled (1, 1, 0, 1).

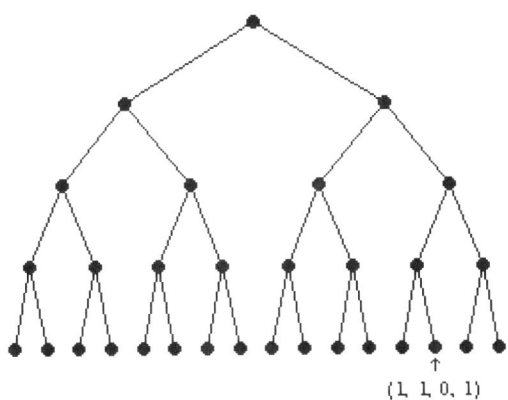

(1, 1, 0, 1)

33. The path representing (1, 1, 1, 0, 0, 0) is the unique path beginning at the root vertex and ending with the vertex labeled (1, 1, 1, 0, 0, 0).

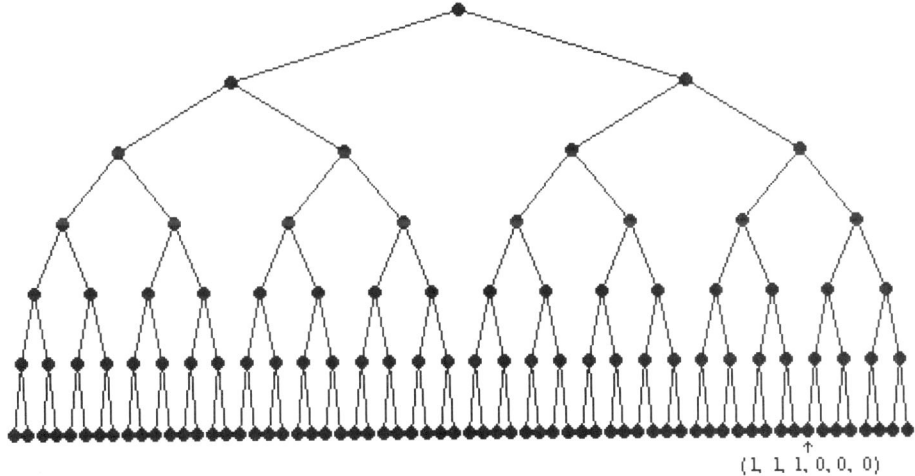

(1, 1, 1, 0, 0, 0)

35.

∅	{1}	{1,2}	{1,2,3}	{1,2,3,4}
	{2}	{1,3}	{1,2,4}	
	{3}	{1,4}	{1,3,4}	
	{4}	{2,3}	{2,3,4}	
		{2,4}		
		{3,4}		

37. (0, 0, 0)   (1, 0, 0)   (1, 1, 0)   (1, 1, 1)
                 (0, 1, 0)   (1, 0, 1)

8   Chapter 1 Sets, Subsets, Induction and Recursion

(0, 0, 1)   (0, 1, 1)

39.   (0, 0, 0, 0, 0)   (1, 0, 0, 0, 0)   (1, 1, 0, 0, 0)   (1, 1, 1, 0, 0)   (1, 1, 1, 1, 0)   (1, 1, 1, 1, 1)
       (0, 1, 0, 0, 0)   (1, 0, 1, 0, 0)   (1, 1, 0, 1, 0)   (1, 1, 1, 0, 1)
       (0, 0, 1, 0, 0)   (1, 0, 0, 1, 0)   (1, 1, 0, 0, 1)   (1, 1, 0, 1, 1)
       (0, 0, 0, 1, 0)   (1, 0, 0, 0, 1)   (1, 0, 1, 1, 0)   (1, 0, 1, 1, 1)
       (0, 0, 0, 0, 1)   (0, 1, 1, 0, 0)   (1, 0, 1, 0, 1)   (0, 1, 1, 1, 1)
                          (0, 1, 0, 1, 0)   (1, 0, 0, 1, 1)
                          (0, 1, 0, 0, 1)   (0, 1, 1, 1, 0)
                          (0, 0, 1, 1, 0)   (0, 1, 1, 0, 1)
                          (0, 0, 1, 0, 1)   (0, 1, 0, 1, 1)
                          (0, 0, 0, 1, 1)   (0, 0, 1, 1, 1)

41. The Standard Gray Code listing of binary vectors of length six is given below where the list is read from top to bottom and left to right.

(0, 0, 0, 0, 0, 0)	(0, 1, 1, 0, 0, 0)	(1, 1, 0, 0, 0, 0)	(1, 0, 1, 0, 0, 0)
(0, 0, 0, 0, 0, 1)	(0, 1, 1, 0, 0, 1)	(1, 1, 0, 0, 0, 1)	(1, 0, 1, 0, 0, 1)
(0, 0, 0, 0, 1, 1)	(0, 1, 1, 0, 1, 1)	(1, 1, 0, 0, 1, 1)	(1, 0, 1, 0, 1, 1)
(0, 0, 0, 0, 1, 0)	(0, 1, 1, 0, 1, 0)	(1, 1, 0, 0, 1, 0)	(1, 0, 1, 0, 1, 0)
(0, 0, 0, 1, 1, 0)	(0, 1, 1, 1, 1, 0)	(1, 1, 0, 1, 1, 0)	(1, 0, 1, 1, 1, 0)
(0, 0, 0, 1, 1, 1)	(0, 1, 1, 1, 1, 1)	(1, 1, 0, 1, 1, 1)	(1, 0, 1, 1, 1, 1)
(0, 0, 0, 1, 0, 1)	(0, 1, 1, 1, 0, 1)	(1, 1, 0, 1, 0, 1)	(1, 0, 1, 1, 0, 1)
(0, 0, 0, 1, 0, 0)	(0, 1, 1, 1, 0, 0)	(1, 1, 0, 1, 0, 0)	(1, 0, 1, 1, 0, 0)
(0, 0, 1, 1, 0, 0)	(0, 1, 0, 1, 0, 0)	(1, 1, 1, 1, 0, 0)	(1, 0, 0, 1, 0, 0)
(0, 0, 1, 1, 0, 1)	(0, 1, 0, 1, 0, 1)	(1, 1, 1, 1, 0, 1)	(1, 0, 0, 1, 0, 1)
(0, 0, 1, 1, 1, 1)	(0, 1, 0, 1, 1, 1)	(1, 1, 1, 1, 1, 1)	(1, 0, 0, 1, 1, 1)
(0, 0, 1, 1, 1, 0)	(0, 1, 0, 1, 1, 0)	(1, 1, 1, 1, 1, 0)	(1, 0, 0, 1, 1, 0)
(0, 0, 1, 0, 1, 0)	(0, 1, 0, 0, 1, 0)	(1, 1, 1, 0, 1, 0)	(1, 0, 0, 0, 1, 0)
(0, 0, 1, 0, 1, 1)	(0, 1, 0, 0, 1, 1)	(1, 1, 1, 0, 1, 1)	(1, 0, 0, 0, 1, 1)
(0, 0, 1, 0, 0, 1)	(0, 1, 0, 0, 0, 1)	(1, 1, 1, 0, 0, 1)	(1, 0, 0, 0, 0, 1)
(0, 0, 1, 0, 0, 0)	(0, 1, 0, 0, 0, 0)	(1, 1, 1, 0, 0, 0)	(1, 0, 0, 0, 0, 0)

Section 1.3 Advanced Exercises

In each exercise, the appropriate Standard Gray Code is given along with the corresponding knapsack weight. The solution to the knapsack problem, the set of objects having maximum weight $\leq N$ is highlighted.

1. $N = 32$

(0, 0, 0, 0)	0	(1, 1, 0, 0)	13
(0, 0, 0, 1)	18	(1, 1, 0, 1)	**31**
(0, 0, 1, 1)	34	(1, 1, 1, 1)	47
(0, 0, 1, 0)	16	(1, 1, 1, 0)	29
(0, 1, 1, 0)	23	(1, 0, 1, 0)	22
(0, 1, 1, 1)	41	(1, 0, 1, 1)	40
(0, 1, 0, 1)	25	(1, 0, 0, 1)	24

(0, 1, 0, 0)	7	(1, 0, 0, 0)	6

3. $N = 66$

(0, 0, 0, 0)	0	(1, 1, 0, 0)	30
(0, 0, 0, 1)	28	(1, 1, 0, 1)	58
(0, 0, 1, 1)	49	(1, 1, 1, 1)	79
(0, 0, 1, 0)	21	(1, 1, 1, 0)	51
(0, 1, 1, 0)	39	(1, 0, 1, 0)	33
(0, 1, 1, 1)	67	(1, 0, 1, 1)	61
(0, 1, 0, 1)	46	(1, 0, 0, 1)	40
(0, 1, 0, 0)	18	(1, 0, 0, 0)	12

5. $N = 54$

(0, 0, 0, 0, 0)	0	(0, 1, 1, 0, 0)	30	(1, 1, 0, 0, 0)	21	(1, 0, 1, 0, 0)	23
(0, 0, 0, 0, 1)	25	(0, 1, 1, 0, 1)	55	(1, 1, 0, 0, 1)	46	(1, 0, 1, 0, 1)	48
(0, 0, 0, 1, 1)	44	(0, 1, 1, 1, 1)	74	(1, 1, 0, 1, 1)	65	(1, 0, 1, 1, 1)	67
(0, 0, 0, 1, 0)	19	(0, 1, 1, 1, 0)	49	(1, 1, 0, 1, 0)	40	(1, 0, 1, 1, 0)	42
(0, 0, 1, 1, 0)	35	(0, 1, 0, 1, 0)	33	(1, 1, 1, 1, 0)	56	(1, 0, 0, 1, 0)	26
(0, 0, 1, 1, 1)	60	(0, 1, 0, 1, 1)	58	(1, 1, 1, 1, 1)	81	(1, 0, 0, 1, 1)	51
(0, 0, 1, 0, 1)	41	(0, 1, 0, 0, 1)	39	(1, 1, 1, 0, 1)	62	(1, 0, 0, 0, 1)	32
(0, 0, 1, 0, 0)	16	(0, 1, 0, 0, 0)	14	(1, 1, 1, 0, 0)	37	(1, 0, 0, 0, 0)	7

10   Chapter 1 Sets, Subsets, Induction and Recursion

Section 1.4 Exercises

1. The set $(A \cup B) - C$ consists of the elements in either A or B that are not in C.

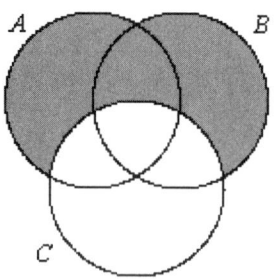

3. The set $(A - B) - C$ consists of those elements of A not in B or C.

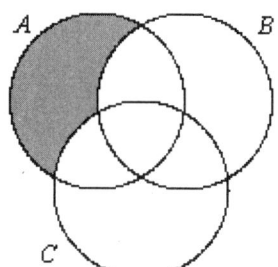

5. The set $A \cap (B \cup C)$ consists of those elements of A also in B or C or both.

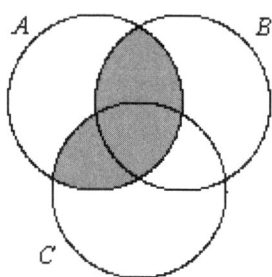

7. The set $A \cap (B \Delta C)$ consists of those elements of A that are also in one of B or C but not both.

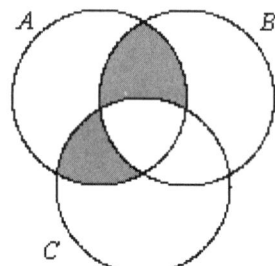

In exercises 9-19 an argument using elements is employed to prove the given set identity. This argument is followed by a pair of equivalent Venn diagrams corresponding to the two sides of the set identity.

9. Let $x \in A \cup (B \cup C)$. If $x \in A$, then $x \in A \cup B$ and so $x \in (A \cup B) \cup C$. If $x \in B \cup C$, then $x \in B$ or $x \in C$. If $x \in B$ then $x \in A \cup B$ and so $x \in (A \cup B) \cup C$. If $x \in C$ then $x \in (A \cup B) \cup C$. Thus $A \cup (B \cup C) \subset (A \cup B) \cup C$. The argument in the other direction is exactly similar.

Shading $A \cup (B \cup C)$ or $(A \cup B) \cup C$ in a Venn diagram results in all three sets being shaded.

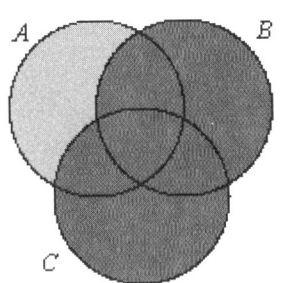 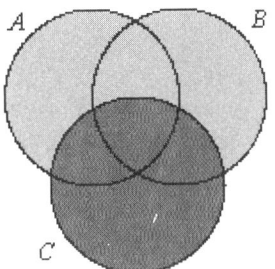

11. Let $x \in A \cap (B \Delta C)$. Then either $x \in A$ and $x \in B \Delta C$ so x is in B or C but not both. If $x \in B$ and $x \notin C$ then $x \in A \cap B$ and $x \notin A \cap C$ meaning $x \in (A \cap B) \Delta (A \cap C)$. The result is the same if $x \notin B$ and $x \in C$. Thus $A \cap (B \Delta C) \subset (A \cap B) \Delta (A \cap C)$.

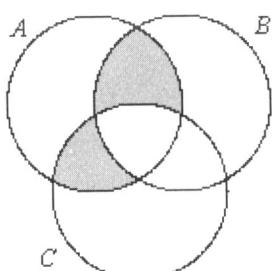 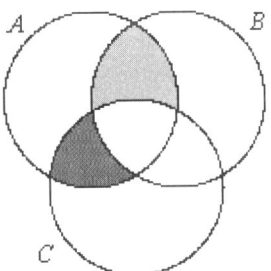

13. For any $x \in X$, x is either in the set A or it is not in A. That is $x \in A$ or $x \in A^c$ i.e. $x \in A \cup A^c$. This shows $X \subset A \cup A^c$. We must have $A \cup A^c \subset X$ since X contains everything. Thus $A \cup A^c = X$.

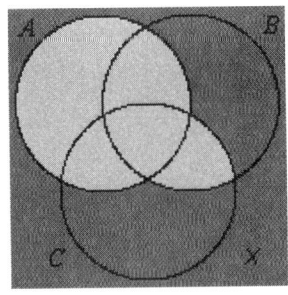 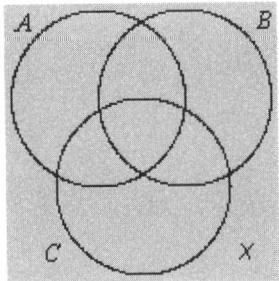

15. Suppose $x \in (A \cap B \cap C)^c$ then x is not in the intersection of A, B and C. There must be at least one of the sets, say A, which does not contain x, meaning $x \in A^c$ and so $x \in A^c \cup B^c \cup C^c$. Similarly if x is not in B or not in C, proving $(A \cap B \cap C)^c \subset A^c \cup B^c \cup C^c$. Conversely, if $x \in A^c \cup B^c \cup C^c$ then there is at least one of the sets A, B and C for whom x is a member of the complement and therefore $x \notin A \cap B \cap C$ or $x \in (A \cap B \cap C)^c$ giving $A^c \cup B^c \cup C^c \subset (A \cap B \cap C)^c$ and proving $A^c \cup B^c \cup C^c = (A \cap B \cap C)^c$.

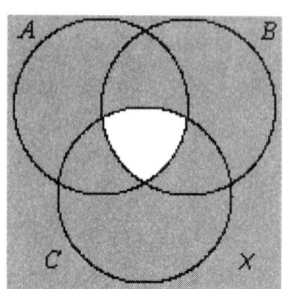
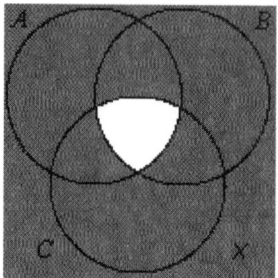

17. The set $A \Delta B$ is the set of elements of X that are in exactly one of the sets A and B. Thus $x \in (A \Delta B)^c$ if and only if x is in neither set ($x \notin A \cup B$ or $x \in (A \cup B)^c$) or x is in both sets ($x \in A \cap B$). Hence
$(A \Delta B)^c = (A \cup B)^c \cup (A \cap B)$

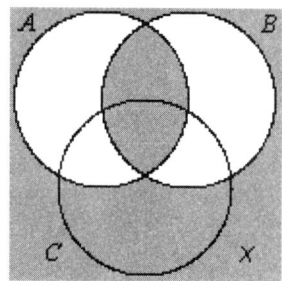
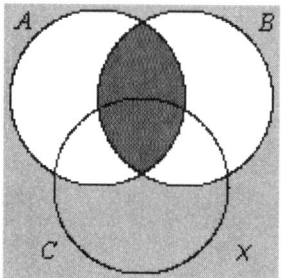

19. If $x \in (A \cap B) - (B \cap C)$ then $x \in A \cap B$ and $x \notin B \cap C$ which implies $x \notin C$ since $x \in B$. Thus $x \in (A \cap B) - C$ and we have shown $(A \cap B) - (B \cap C) \subset (A \cap B) - C$. If $x \in (A \cap B) - C$ then $x \in A \cap B$ and $x \notin C$. Since $x \notin C$ we have $x \notin B \cap C$ giving $x \in (A \cap B) - C$ and proving $(A \cap B) - C \subset (A \cap B) - (B \cap C)$ and $(A \cap B) - (B \cap C) = (A \cap B) - C$.

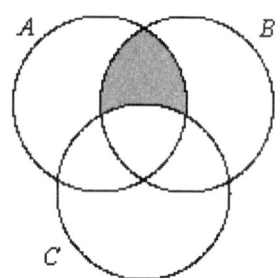
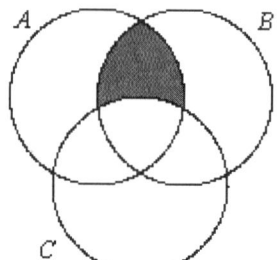

21. By the Principle of Inclusion-Exclusion,
$$|A_1 \cup A_2 \cup A_3| = |A_1| + |A_2| + |A_3| - |A_1 \cap A_2| - |A_1 \cap A_3| - |A_2 \cap A_3| + |A_1 \cap A_2 \cap A_3|$$
$$= 12 + 12 + 12 - 8 - 8 - 8 + 6 = 18.$$

23. By the Principle of Inclusion-Exclusion,
$$|A_1 \cup A_2 \cup A_3| = |A_1| + |A_2| + |A_3| - |A_1 \cap A_2| - |A_1 \cap A_3| - |A_2 \cap A_3| + |A_1 \cap A_2 \cap A_3|$$
$$= 5 + 5 + 5 - 1 - 3 - 1 + 1 = 11.$$

25. With the notation of Exercise 24, we first compute $|A_1 \cap A_2| + |A_1 \cap A_3| + |A_2 \cap A_3| = 24$. This sum counts the townspeople in all three clubs $A_1 \cap A_2 \cap A_3$ three times so we subtract the two extras to find $24 - 2(6) = 12$ townspeople belong to two or more clubs.

27. $(1, 0, 0, 0, 1, 1, 0, 1) + (1, 0, 1, 1, 0, 1, 0, 0) = (0, 0, 1, 1, 1, 0, 0, 1)$ which is in H_8.

Section 1.4 Advanced Exercises

1.

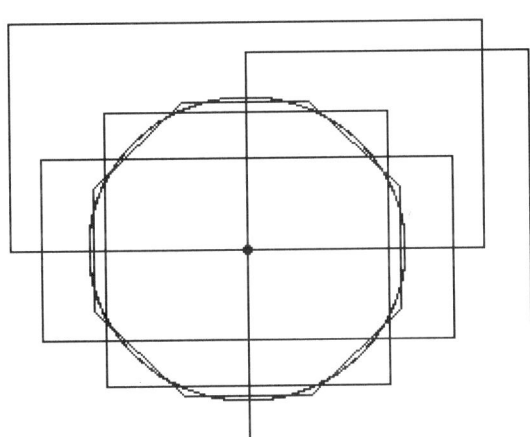

14 Chapter 1 Sets, Subsets, Induction and Recursion

3.

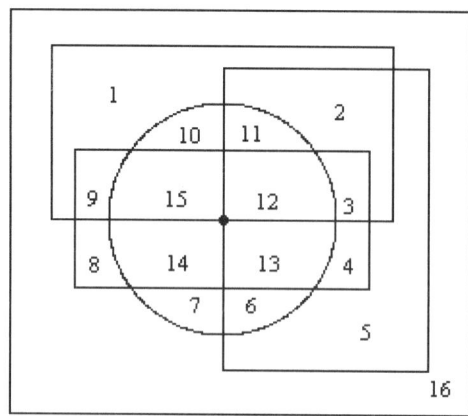

5. Starting with (0,0) and given that the next element in any Gray code must differ from (0,0) in one place, there are two choices for the second element of a Gray code, namely (1,0) and (0,1). Thus any Gray codes for binary vectors of length two must begin with either (0, 0), (1, 0) or (0, 0), (0, 1). In either case, the third vector cannot differ from the second in the 1's position since (0, 0) is already on the list. Therefore (1, 1) must be the third vector in any Gray codes, leaving only one possibility for the fourth position. The only Gray codes for vectors of length two are

(0, 0)	(0, 0)
(1, 0)	(0, 1)
(1, 1)	(1, 1)
(0, 1)	(1, 0) .

Section 1.5 Exercises

1. Beginning with $F_0 = 0$ and $F_1 = 1$ and applying the Fibonacci recursion $F_n = F_{n-1} + F_{n-2}$ for $n = 2$ to 20 yields the sequence 0, 1, 1, 2, 3, 5, 8, 13, 21, 34, 55, 89, 144, 233, 377, 610, 987, 1597, 2584, 4181, 6765.

3. Beginning with $T_0 = 0$, $T_1 = 0$, $T_2 = 1$ and applying the Tribonacci recursion $T_n = T_{n-1} + T_{n-2} + T_{n-3}$ for $n = 3$ to 20 yields the sequence 0, 0, 1, 1, 2, 4, 7, 13, 24, 44, 81, 149, 274, 504, 927, 1705, 3136, 5768, 10609, 19513, 35890.

4. From the sixth row of the Pascal triangle, we construct the relation $B_0 + 6B_1 + 15B_2 + 20B_3 + 15B_4 + 6B_5 = 0$. Substituting $B_0 = 1$, $B_1 = -\frac{1}{2}$, $B_2 = \frac{1}{6}$, $B_3 = 0, B0 = 1$, and $B_4 = -\frac{1}{30}$ as derived in Example 4, gives $6B_5 = 0$ or $B_5 = 0$.

5. From the seventh row of the Pascal triangle, we construct the relation $B_0 + 7B_1 + 21B_2 + 35B_3 + 35B_4 + 21B_5 + 7B_6 = 0$. Using the values from Example 4 and the fact that $B_5 = 0$, we find $7B_6 = -\frac{1}{6}$ or $B_6 = -\frac{1}{42}$.

7. From the ninth row of the Pascal triangle, we construct the relation
$B_0 + 9B_1 + 36B_2 + 84B_3 + 126B_4 + 126B_5 + 84B_6 + 36B_7 + 9B_8 = 0$. Using the values from Example 4 and the results $B_5 = 0$, $B_6 = -\frac{1}{42}$, and $B_7 = 0$, we find $9B_8 = \frac{3}{10}$ or $B_8 = \frac{1}{30}$.

9. The Collatz sequence with initial value $C[0] = 19$ is 19, 58, 29, 88, 44, 22, 11, 34, 17, 52, 26, 13, 40, 20, 10, 5, 16, 8, 4, 2, 1.

11. The Collatz sequence with initial value $C[0] = 123$ is 123, 370, 185, 556, 278, 139, 418, 209, 628, 314, 157, 472, 236, 118, 59, 178, 89, 268, 134, 67, 202, 101, 304, 152, 76, 38, 19, 58, 29, 88, 44, 22, 11, 34, 17, 52, 26, 13, 40, 20, 10, 5, 16, 8, 4, 2, 1.

13. The Collatz sequence with initial value $C[0] = 251$ is 251, 754, 377, 1132, 566, 283, 850, 425, 1276, 638, 319, 958, 479, 1438, 719, 2158, 1079, 3238, 1619, 4858, 2429, 7288, 3644, 1822, 911, 2734, 1367, 4102, 2051, 6154, 3077, 9232, 4616, 2308, 1154, 577, 1732, 866, 433, 1300, 650, 325, 976, 488, 244, 122, 61, 184, 92, 46, 23, 70, 35, 106, 53, 160, 80, 40, 20, 10, 5, 16, 8, 4, 2, 1.

15. The Collatz sequence with initial value $C[0] = 103$ is 103, 310, 155, 466, 233, 700, 350, 175, 526, 263, 790, 395, 1186, 593, 1780, 890, 445, 1336, 668, 334, 167, 502, 251, 754, 377, 1132, 566, 283, 850, 425, 1276, 638, 319, 958, 479, 1438, 719, 2158, 1079, 3238, 1619, 4858, 2429, 7288, 3644, 1822, 911, 2734, 1367, 4102, 2051, 6154, 3077, 9232, 4616, 2308, 1154, 577, 1732, 866, 433, 1300, 650, 325, 976, 488, 244, 122, 61, 184, 92, 46, 23, 70, 35, 106, 53, 160, 80, 40, 20, 10, 5, 16, 8, 4, 2, 1.

17. The Collatz sequence with initial value $C[0] = 27$ is 27, 82, 41, 124, 62, 31, 94, 47, 142, 71, 214, 107, 322, 161, 484, 242, 121, 364, 182, 91, 274, 137, 412, 206, 103, 310, 155, 466, 233, 700, 350, 175, 526, 263, 790, 395, 1186, 593, 1780, 890, 445, 1336, 668, 334, 167, 502, 251, 754, 377, 1132, 566, 283, 850, 425, 1276, 638, 319, 958, 479, 1438, 719, 2158, 1079, 3238, 1619, 4858, 2429, 7288, 3644, 1822, 911, 2734, 1367, 4102, 2051, 6154, 3077, 9232, 4616, 2308, 1154, 577, 1732, 866, 433, 1300, 650, 325, 976, 488, 244, 122, 61, 184, 92, 46, 23, 70, 35, 106, 53, 160, 80, 40, 20, 10, 5, 16, 8, 4, 2, 1.

19. The Fibonacci tree of seven levels is shown below. Note that when building the Fibonacci tree, an edge is considered a right edge only if there is an edge to its left off the same node. Otherwise the edge is considered a left edge even if it is drawn to the right.

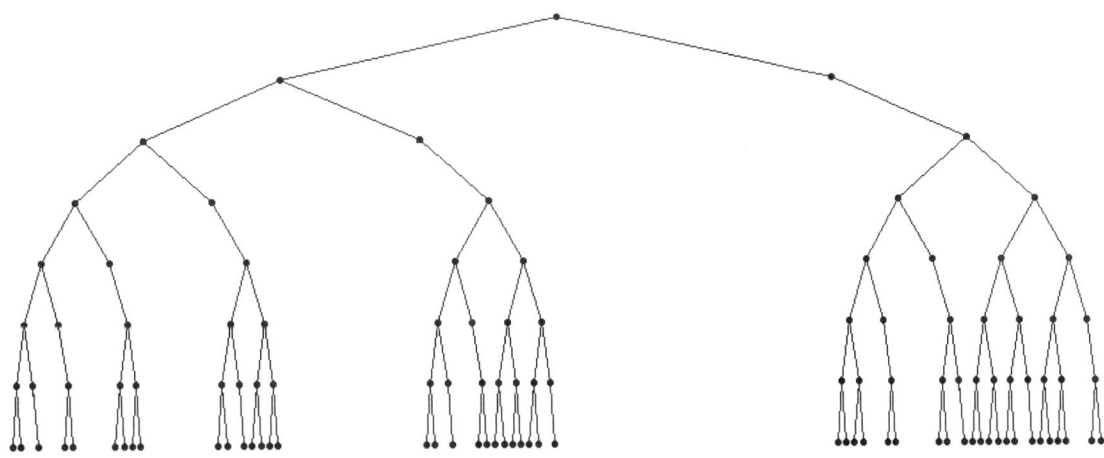

21. Let S_n denote the number of ways to climb n stairs taken one, two or three at a time. By an identical argument to that used in Exercise 21, we have $S_n = S_{n-1} + S_{n-2} + S_{n-3}$ with $S_1 = 1$, $S_2 = 2$ and $S_3 = 4$ (one stair

16   Chapter 1 Sets, Subsets, Induction and Recursion

at a time, two stairs then one, one stair then two, all three stairs at once.) This is the Tribonacci sequence shifted by two so $S_n = T_{n+2}$.

In Exercises 23-27, the appropriate difference table is given. The steps from the last row to the first that produce the next term of the sequence are highlighted.

23.

	2		5		10		17		26		37		50		65
Δ		3		5		7		9		11		13		15	
Δ^2			2		2		2		2		2		2		

25.

	1		4		11		22		37		56		79		106
Δ		3		7		11		15		19		23		27	
Δ^2			4		4		4		4		4		4		

27.

	1		7		25		61		121		211		337		505
Δ		6		18		36		60		90		126		168	
Δ^2			12		18		24		30		36		42		
Δ^3				6		6		6		6		6			

29. Iterating the recursion formula $a_n = a_{n-1} + a_{n-3}$ gives the sequence of values
0, 0, 1, 1, 1, 0, 1, 0, 0, 1, …Note the initial values have now repeated so the sequence consists of the subsequence 0, 0, 1, 1, 1, 0, 1 repeating indefinitely and has period 7.

Section 1.5 Advanced Exercises

1.

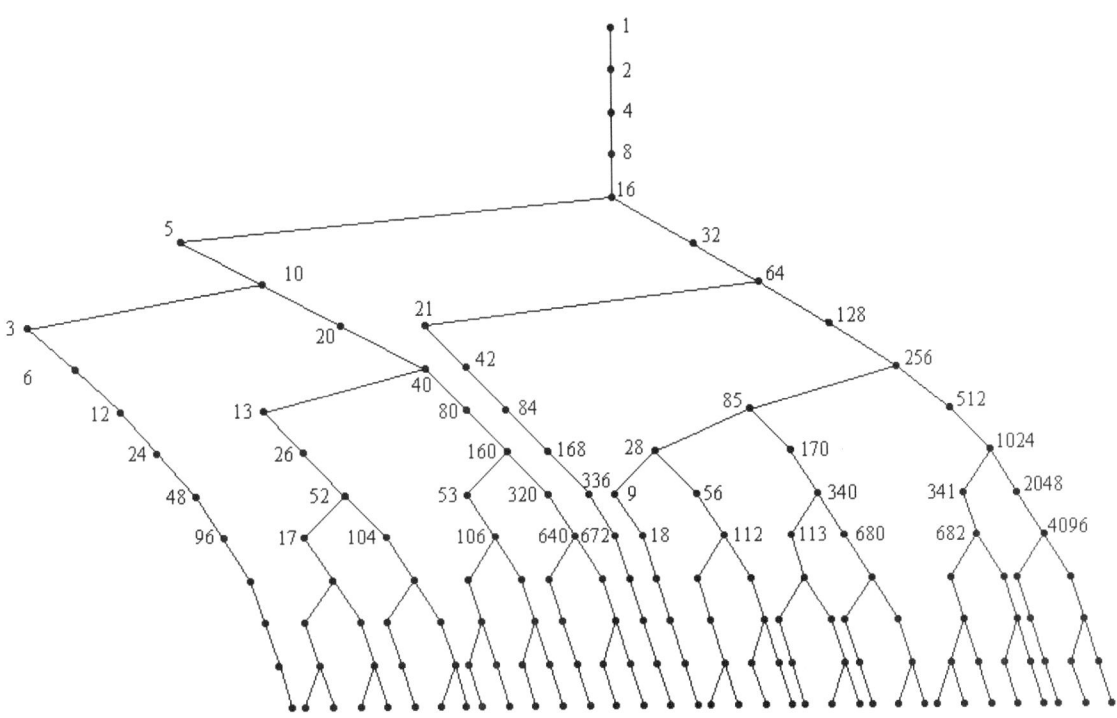

The level sequence is 1, 1, 1, 1, 1, 2, 2, 4, 4, 6, 7, 10, 12, 17, 21, 28, 35.

Section 1.6 Supplementary Exercises

1. $\binom{6}{3} = 20$ 3. $\binom{8}{4} = 70$ 5. $\binom{12}{7} = 792$

7. $\sum_{k=1}^{1}(5k-4) = 1 = \frac{1 \cdot (5 \cdot 1 - 3)}{2}$ providing the initial step in the induction argument.

Now suppose $\sum_{k=1}^{n}(5k-4) = \frac{n(5n-3)}{2}$ for $n > 1$. Then

$$\sum_{k=1}^{n+1}(5k-4) = \sum_{k=1}^{n}(5k-4) + (5(n+1)-4) = \frac{n(5n-3)}{2} + (5n+1) = \frac{5n^2+7n+2}{2} = \frac{(n+1)(5(n+1)-3)}{2}$$

giving the induction step and proving the identity $\sum_{k=1}^{n}(5k-4) = \frac{n(5n-3)}{2}$.

9. $\sum_{k=1}^{1}\frac{1}{4k^2-1} = \frac{1}{3} = \frac{1}{2 \cdot 1 + 1}$ providing the initial step in the induction argument.

18   Chapter 1   Sets, Subsets, Induction and Recursion

Now suppose $\sum_{k=1}^{n}\dfrac{1}{4k^2-1}=\dfrac{n}{2n+1}$ for $n>1$. Then

$$\sum_{k=1}^{n+1}\dfrac{1}{4k^2-1}=\sum_{k=1}^{n}\dfrac{1}{4k^2-1}+\dfrac{1}{4(n+1)^2-1}=\dfrac{n}{2n+1}+\dfrac{1}{4(n+1)^2-1}=\dfrac{2n^2+3n+1}{4n^2+8n+3}=\dfrac{n+1}{2(n+1)+1}$$

giving the induction step and proving the identity $\sum_{k=1}^{n}\dfrac{1}{4k^2-1}=\dfrac{n}{2n+1}$.

11. $\{a \in N : a = 8b+7 \text{ for some } b \in N \cup \{0\}\}$

13. $A \times B = \{(2,1),(2,4),(2,7),(2,9),(6,1),(6,4),(6,7),(6,9),(7,1),(7,4),(7,7),(7,9),$
    $(8,1),(8,4),(8,7),(8,9),(9,1),(9,4),(9,7),(9,9),\}$
    $A \cup B = \{1,2,4,6,7,8,9\}$
    $A \cap B = \{7,9\}$
    $A - B = \{2,6,8\}$
    $A^c = \{1,3,4,5\}$
    $B^c = \{2,3,5,6,8\}$

15. The path representing {1, 3} is indicated by the heavy line in the tree below.

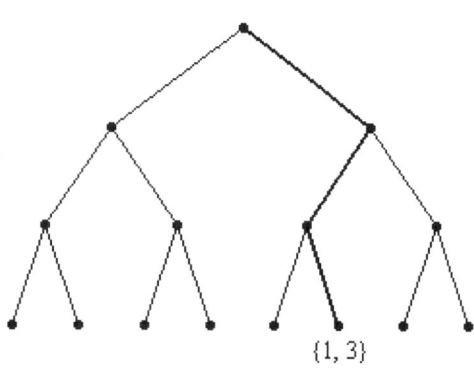

{1, 3}

17.

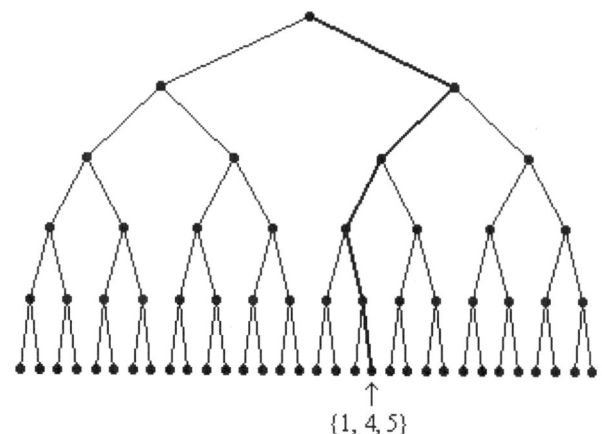

{1, 4, 5}

19. A listing of the Standard Gray code can be found in the solution to Exercise 41 of Section 1.3. The corresponding subsets of {1, 2, 3, 4, 5, 6} are listed below.

∅	{2, 3}	{1, 2}	{1, 3}
{6}	{2, 3, 6}	{1, 2, 6}	{1, 3, 6}
{5, 6}	{2, 3, 5, 6}	{1, 2, 5, 6}	{1, 3, 5, 6}
{5}	{2, 3, 5}	{1, 2, 5}	{1, 3, 5}
{4, 5}	{2, 3, 4, 5}	{1, 2, 4, 5}	{1, 3, 4, 5}
{4, 5, 6}	{2, 3, 4, 5, 6}	{1, 2, 4, 5, 6}	{1, 3, 4, 5, 6}
{4, 6}	{2, 3, 4, 6}	{1, 2, 4, 6}	{1, 3, 4, 6}
{4}	{2, 3, 4}	{1, 2, 4}	{1, 3, 4}
{3, 4}	{2, 4}	{1, 2, 3, 4}	{1, 4}
{3, 4, 6}	{2, 4, 6}	{1, 2, 3, 4, 6}	{1, 4, 6}
{3, 4, 5, 6}	{2, 4, 5, 6}	{1, 2, 3, 4, 5, 6}	{1, 4, 5, 6}
{3, 4, 5}	{2, 4, 5}	{1, 2, 3, 4, 5}	{1, 4, 5}
{3, 5}	{2, 5}	{1, 2, 3, 5}	{1, 5}
{3, 5, 6}	{2, 5, 6}	{1, 2, 3, 5, 6}	{1, 5, 6}
{3, 6}	{2, 6}	{1, 2, 3, 6}	{1, 6}
{3}	{2}	{1, 2, 3}	{1}

21. For any $x \in A$, either x is in the set B or in B^c. If $x \in B$ then $x \in A \cap B$ and so $x \in (A \cap B) \cup (A \cap B^c)$. Similarly if $x \in B^c$ it follows that $x \in (A \cap B) \cup (A \cap B^c)$ and so $A \subset (A \cap B) \cup (A \cap B^c)$. Now note each of the sets $A \cap B$ and $A \cap B^c$ are subsets of A and so $(A \cap B) \cup (A \cap B^c) \subset A$ proving $A = (A \cap B) \cup (A \cap B^c)$.

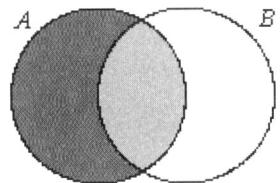

23. $|A_1 \cup A_2 \cup A_3| = |A_1| + |A_2| + |A_3| - |A_1 \cap A_2| - |A_1 \cap A_3| - |A_2 \cap A_3| + |A_1 \cap A_2 \cap A_3|$
$= 4 + 4 + 4 - 0 - 0 - 3 + 0 = 9$.

25. $|A_1 \cup A_2 \cup A_3| = |A_1| + |A_2| + |A_3| - |A_1 \cap A_2| - |A_1 \cap A_3| - |A_2 \cap A_3| + |A_1 \cap A_2 \cap A_3|$
$= 6 + 5 + 6 - 4 - 4 - 4 + 4 = 9$.

27. $\{1, 3, 7, 8\} \Delta \{1, 2, 6, 7\} = \{2, 3, 6, 8\}$ which is in H_8.

29.

	0		3		8		15		24		35		48		63
Δ		3		5		7		9		11		13		15	
Δ^2			2		2		2		2		2		2		

31.

		-1		-2		3		20		55		114		203		328
Δ			-1		5		17		35		59		89		125	
Δ^2				6		12		18		24		30		36		
Δ^3					6		6		6		6		6			

Section 1.6 Advanced Exercises

1.

$$F_1 = \binom{1}{0} = 1$$

$$F_2 = \binom{2}{0} + \binom{1}{1} = 1 + 1 = 2$$

$$F_3 = \binom{3}{0} + \binom{2}{1} = 1 + 2 = 3$$

$$F_4 = \binom{4}{0} + \binom{3}{1} + \binom{2}{2} = 1 + 3 + 1 = 5$$

$$F_5 = \binom{5}{0} + \binom{4}{1} + \binom{3}{2} = 1 + 4 + 3 = 8$$

$$F_6 = \binom{6}{0} + \binom{5}{1} + \binom{4}{2} + \binom{3}{3} = 1 + 5 + 6 + 1 = 13$$

$$F_7 = \binom{7}{0} + \binom{6}{1} + \binom{5}{2} + \binom{4}{3} = 1 + 6 + 10 + 4 = 21$$

$$F_8 = \binom{8}{0} + \binom{7}{1} + \binom{6}{2} + \binom{5}{3} + \binom{4}{4} = 1 + 7 + 15 + 10 + 1 = 34$$

$$F_9 = \binom{9}{0} + \binom{8}{1} + \binom{7}{2} + \binom{6}{3} + \binom{5}{4} = 1 + 8 + 21 + 20 + 5 = 55$$

$$F_{10} = \binom{10}{0} + \binom{9}{1} + \binom{8}{2} + \binom{7}{3} + \binom{6}{4} + \binom{5}{5} = 1 + 9 + 28 + 35 + 15 + 6 = 89$$

$$F_{11} = \binom{11}{0} + \binom{10}{1} + \binom{9}{2} + \binom{8}{3} + \binom{7}{4} + \binom{6}{5} = 1 + 10 + 36 + 56 + 35 + 6 = 144$$

$$F_{12} = \binom{12}{0} + \binom{11}{1} + \binom{10}{2} + \binom{9}{3} + \binom{8}{4} + \binom{7}{5} + \binom{6}{6} = 1 + 11 + 45 + 84 + 70 + 21 + 1 = 233$$

3. The rows of this triangle can be defined recursively as follows. The first row, row 1, consists of the single entry 1. Thereafter, row n contains n entries, the first of which is 1 and the last of which is 2. The entries in between 1 and 2 alternate between being equal to the value in the row above to the left of the entry and being equal to the sum of the values to the left and right in the row above.

The sum of elements in the first three rows are 1, 3, 4 equal to L_1, L_2, and L_3, the first three Lucas numbers. Now suppose the entries in the kth row for $k < n$ sum to L_k. Denote the entries in rows $n-2$, $n-1$, and n by a_i, b_i and c_i respectively for $i = 1, 2, \ldots$. The recursive definition of the triangle states $a_1 = b_1 = c_1 = 1$, $a_{n-2} = b_{n-1} = c_n = 2$, and $c_2 = b_1$, $c_3 = b_2 + b_3$, $c_4 = b_3$ and so on. Similar formulas hold for the b_i's in terms of the a_i's. It follows then that

$$
\begin{aligned}
c_1 + c_2 + c_3 + \ldots &= b_1 + b_1 + (b_2 + b_3) + b_3 + (b_4 + b_5) + b_5 + \ldots = (b_1 + b_2 + b_3 + b_4 + b_5 + \ldots) + (b_1 + b_3 + b_5 + \ldots) \\
&= (b_1 + b_2 + b_3 + b_4 + b_5 + \ldots) + (a_1 + (a_2 + a_3) + (a_4 + a_5) + \ldots) \\
&= L_{n-1} + L_{n-2} \\
&= L_n
\end{aligned}
$$

where the second to last equality is justified by the induction hypothesis and the last equality follows from the definition of the Lucas numbers. This completes the induction argument, proving the sum of the entries in row n is L_n.

Chapter 2 Integers, Remainders, and the Golden Ratio

Section 2.1 Exercises

1. Applying the Euclidean algorithm,
    $64 = 2 \cdot 28 + 8$
    $28 = 3 \cdot 8 + 4$ so g.c.d. $(64,28) = 4$
   l.c.m. $(64, 28) = (64 \cdot 28)/$ g.c.d. $(64,28) = 64 \cdot 28 / 4 = 448$
   Lastly $4 = 28 - 3 \cdot 8 = 28 - 3 \cdot (64 - 2 \cdot 28) = -3 \cdot 64 + 7 \cdot 28$

3. g.c.d.$(130, 23) = 1 = -3 \cdot 130 + 17 \cdot 23$ l.c.m.$(130, 23) = 2990$

5. g.c.d.$(6765, 4181) = 1 = 1597 \cdot 6765 - 2584 \cdot 4181$ l.c.m.$(130, 23) = 28284465$

7. The Euclidean Algorithm requires 14 steps to compute g.c.d.$(1597, 987) = 1$ while g.c.d.$(1590, 997) = 1$ requires 8 steps. The reason for the difference is the fact that 987 and 107 are consecutive Fibonacci numbers as revealed by Lamé's Theorem in Section 2.2.

9. g.c.d.$(93, 47) = 1$ in 2 steps.
11. g.c.d.$(605, 322) = 1$ in 5 steps.
13. g.c.d.$(70000, 38502) = 2$ in 5 steps.
15. $95,040 = 2^6 \cdot 3^3 \cdot 5 \cdot 11$
17. $10,200,960 = 2^7 \cdot 3^2 \cdot 5 \cdot 7 \cdot 11 \cdot 23$

19. We seek integers x and y such that $y^2 = x^2 - 2183$. Beginning with $x = 47 \geq \sqrt{2183} \approx 46.723$ and incrementing x by 1 until $y = \sqrt{x^2 - 2183}$ is an integer we find $x = 48$ and $y = 11$. Thus $2183 = (x - y)(x + y) = (37) \cdot (59)$.

21. $x = 130$, $y = 21$, $16459 = (109) \cdot (151)$
23. $x = 339$, $y = 58$, $111557 = (281) \cdot (397)$

25.

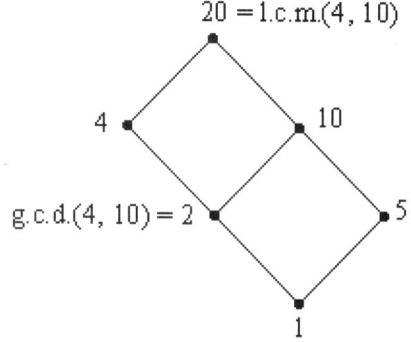

27. The graph Div(105) is shown below. g.c.d.(15,21) = 3 and l.c.m.(15,21) = 105.

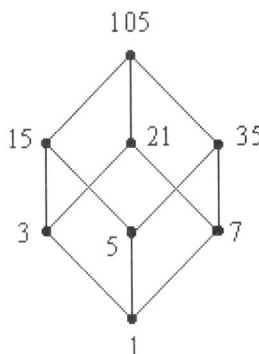

29 – 31. The graph Div(210) is shown below. In Exercise 29, g.c.d.(6,15) = 3 and l.c.m.(6,15) = 30. In Exercise 31, g.c.d.(6,35) = 1 and l.c.m.(6,35) = 210.

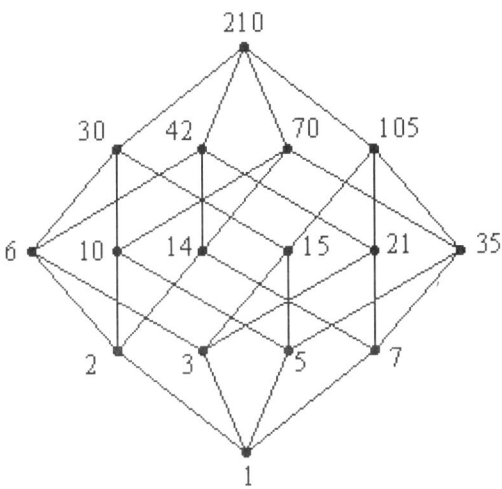

Section 2.1 Advanced Exercises

1. We wish to show $2^n - 1$ is not prime if the integer n is not prime. If n is not prime then $n = a \cdot b$ for some integers, $a > 1$, $b > 1$. Letting $x = 2^a$ in the identity $x^b - 1 = (x-1)(x^{b-1} + x^{b-2} + ... + x + 1)$ gives $2^n - 1 = (2^a)^b - 1 = (2^a - 1)(2^{a(b-1)} + 2^{a(b-2)} + ... + 2^a + 1)$. Since both a and b are greater than 1, the two factors on the right in the last equation above will not equal 1. Thus we have provided a non-trivial factorization of $2^n - 1$ into integers and so $2^n - 1$ cannot be prime.

3. Trial and error finds that $74k + 1$ divides $2^{37} - 1$ when $k = 3$ providing the factorization $2^{37} - 1 = (223)(616318177)$.

5. We seek divisors of $2^{29} - 1$ having the form $58k + 1$. Trial and error shows $k = 4$ gives a divisor of 233 and the factorization $2^{29} - 1 = (233)(2304167)$.

24 Chapter 2 Integers, Remainders, and the Golden Ratio

Section 2.2 Exercises

1. F_{10} = nearest integer to $\dfrac{1}{\sqrt{5}}\left(\dfrac{1+\sqrt{5}}{2}\right)^{10} = 55.004$ or 55.

   L_{10} = nearest integer to $\left(\dfrac{1+\sqrt{5}}{2}\right)^{10} = 122.992$ or 123.

3. F_{40} = nearest integer to $\dfrac{1}{\sqrt{5}}\left(\dfrac{1+\sqrt{5}}{2}\right)^{40} = 102334155.0000001$ or 102334155.

   L_{40} = nearest integer to $\left(\dfrac{1+\sqrt{5}}{2}\right)^{40} = 228826127.0000003$ or 228826127.

5. Applying the Euclidean algorithm to find g.c.d.(988, 602) takes 8 steps

   $988 = 1 \cdot 602 + 386 \quad 602 = 1 \cdot 386 + 216 \quad 386 = 1 \cdot 216 + 170$
   $216 = 1 \cdot 170 + 46 \quad 170 = 3 \cdot 46 + 32 \quad 46 = 1 \cdot 32 + 14$
   $32 = 2 \cdot 14 + 4 \quad 14 = 3 \cdot 4 + 2$

   satisfying the statement of Lamé's Theorem since $8 \leq \dfrac{\ln(602)}{\ln(\phi)} = 13.3003$.

7. Using the Euclidean algorithm to compute g.c.d.(3524580, 2178300) = 60 takes 15 steps validating Lamé's Theorem since $15 \leq \dfrac{\ln(2178300)}{\ln(\phi)} = 30.328$.

9. $F_7 = 2^{-7+1}\left(\binom{7}{1} + \binom{7}{3}\cdot 5 + \binom{7}{5}\cdot 5^2 + \binom{7}{7}\cdot 5^3\right) = \dfrac{1}{64}(7+175+525+125) = 13$.

11. $L_6 = 2^{-6+1}\left(\binom{6}{0} + \binom{6}{2}\cdot 5 + \binom{6}{4}\cdot 5^2 + \binom{6}{6}\cdot 5^3\right) = \dfrac{1}{32}(1+75+375+125) = 18$.

13. $L_8 = 2^{-8+1}\left(\binom{8}{0} + \binom{8}{2}\cdot 5 + \binom{8}{4}\cdot 5^2 + \binom{8}{6}\cdot 5^3 + \binom{8}{8}\cdot 5^4\right) = \dfrac{1}{128}(1+140+1750+3500+625) = 47$.

15. First note $\dfrac{2}{3} \geq \dfrac{1}{2}$ and no other fraction of the form $\dfrac{1}{k}$ is closer to $\dfrac{2}{3}$. Next note $\dfrac{2}{3} - \dfrac{1}{2} = \dfrac{1}{6}$ so the Egyptian Fraction Algorithm gives $\dfrac{2}{3} = \dfrac{1}{2} + \dfrac{1}{6}$.

17. $\dfrac{5}{7} = \dfrac{1}{2} + \dfrac{1}{5} + \dfrac{1}{70}$.

19. $\dfrac{8}{11} = \dfrac{1}{2} + \dfrac{1}{5} + \dfrac{1}{37} + \dfrac{1}{4070}$.

Section 2.2 Advanced Exercises

1. First note $\sum_{k=1}^{1} F_k = F_1 = 1 = 2 - 1 = F_3 - 1$ starting the induction process. Now assume $\sum_{k=1}^{n} F_k = F_{n+2} - 1$ then
$\sum_{k=1}^{n+1} F_k = \sum_{k=1}^{n} F_k + F_{n+1} = F_{n+2} - 1 + F_{n+1} = F_{n+3} - 1$ using the induction hypothesis and the fact that
$F_{n+1} + F_{n+2} = F_{n+3}$. This proves identity I for $n \geq 1$ by induction.

3. Since $F_0 = 0$, $F_1 = 1$, and $F_2 = 1$ it follows that $F_2 F_0 - F_1^2 = -1 = (-1)^1$ and identity III is verified for $n = 1$. Now assume $F_{n+1} F_{n-1} - F_n^2 = (-1)^n$ and consider the quantity $F_{n+2} F_n - F_{n+1}^2$. Substituting the Fibonacci relation $F_{n+2} = F_{n+1} + F_n$ and adding and subtracting the quantity $F_{n+1} F_{n-1}$ we have

$$F_{n+2} F_n - F_{n+1}^2 = (F_{n+1} + F_n) F_n - F_{n+1}^2 - F_{n+1} F_{n-1} + F_{n+1} F_{n-1}$$
$$= (F_n^2 - F_{n+1} F_{n-1}) + F_{n+1} F_{n-1} + F_{n+1} F_n - F_{n+1}^2$$
$$= -(F_{n+1} F_{n-1} - F_n^2) + F_{n+1}(F_{n-1} + F_n - F_{n+1})$$

The first term in the last expression is equal to $(-1)^{n+1}$ by the induction hypothesis. The second term is identically 0 since $F_{n+1} = F_n + F_{n-1}$. This completes the induction step and proves identity III for $n \geq 1$.

5. Using Binet's formulas we have

$$F_{n+1} + F_{n-1} = \frac{1}{\sqrt{5}}\left(\left(\frac{1+\sqrt{5}}{2}\right)^{n+1} - \left(\frac{1-\sqrt{5}}{2}\right)^{n+1}\right) + \frac{1}{\sqrt{5}}\left(\left(\frac{1+\sqrt{5}}{2}\right)^{n-1} - \left(\frac{1-\sqrt{5}}{2}\right)^{n-1}\right)$$
$$= \frac{1}{\sqrt{5}}\left(\left(\frac{1+\sqrt{5}}{2}\right)^{n+1} + \left(\frac{1+\sqrt{5}}{2}\right)^{n-1}\right) - \frac{1}{\sqrt{5}}\left(\left(\frac{1-\sqrt{5}}{2}\right)^{n+1} + \left(\frac{1-\sqrt{5}}{2}\right)^{n-1}\right)$$
$$= \frac{1}{\sqrt{5}}\left(\frac{1+\sqrt{5}}{2}\right)^{n-1}\left(\left(\frac{1+\sqrt{5}}{2}\right)^2 + 1\right) - \frac{1}{\sqrt{5}}\left(\frac{1-\sqrt{5}}{2}\right)^{n-1}\left(\left(\frac{1-\sqrt{5}}{2}\right)^2 + 1\right)$$
$$= \left(\frac{1+\sqrt{5}}{2}\right)^{n-1}\left(\frac{1+\sqrt{5}}{\sqrt{5}}\right) - \left(\frac{1-\sqrt{5}}{2}\right)^{n-1}\left(-\left(\frac{1-\sqrt{5}}{2}\right)\right) = L_n .$$

proving the identity.

7. First note
$$F_k^2 = \frac{1}{5}\left(\left(\frac{1+\sqrt{5}}{2}\right)^{2k} + \left(\frac{1-\sqrt{5}}{2}\right)^{2k} - 2\left(\frac{1+\sqrt{5}}{2}\right)^k\left(\frac{1-\sqrt{5}}{2}\right)^k\right) = \frac{1}{5}\left(\left(\frac{1+\sqrt{5}}{2}\right)^{2k} + \left(\frac{1-\sqrt{5}}{2}\right)^{2k} - 2(-1)^k\right)$$
so

$$F_{n+1}^2 - F_{n-1}^2 = \frac{1}{5}\left(\left(\frac{1+\sqrt{5}}{2}\right)^{2n+2} + \left(\frac{1-\sqrt{5}}{2}\right)^{2n+2} - 2(-1)^{n+1}\right) - \frac{1}{5}\left(\left(\frac{1+\sqrt{5}}{2}\right)^{2n-2} + \left(\frac{1-\sqrt{5}}{2}\right)^{2n-2} - 2(-1)^{n-1}\right)$$

$$= \frac{1}{5}\left(\left(\frac{1+\sqrt{5}}{2}\right)^2 - \left(\frac{1+\sqrt{5}}{2}\right)^{-2}\right)\left(\frac{1+\sqrt{5}}{2}\right)^{2n} + \frac{1}{5}\left(\left(\frac{1-\sqrt{5}}{2}\right)^2 - \left(\frac{1-\sqrt{5}}{2}\right)^{-2}\right)\left(\frac{1-\sqrt{5}}{2}\right)^{2n} - 2\left((-1)^{n+1} - (-1)^{n-1}\right)$$

$$= \frac{1}{\sqrt{5}}\left(\frac{1+\sqrt{5}}{2}\right)^{2n} - \frac{1}{\sqrt{5}}\left(\frac{1-\sqrt{5}}{2}\right)^{2n} - 0 = F_{2n}$$

proving the identity.

9. Fix the integer m and perform induction on n. In the $n = 0$ case
$F_{0+m+1} = F_{m+1} = 1 \cdot F_{m+1} + 0 \cdot F_m = F_1 F_{m+1} + F_0 F_m$ which starts the induction. Now assume
$F_{k+m+1} = F_{k+1} F_{m+1} + F_k F_m$ for $0 \le k < n$. Then
$F_{n+m+1} = F_{n+m} + F_{n+m-1} = (F_n F_{m+1} + F_{n-1} F_m) + (F_{n-1} F_{m+1} + F_{n-2} F_m)$

the last two terms following from the induction hypothesis with $k = n-1$ and $k = n-2$ respectively. By combining terms, factoring and applying the recursion formula for the Fibonacci numbers, we have
$F_{n+m+1} = (F_n + F_{n-1})F_{m+1} + (F_{n-1} + F_{n-2})F_m = F_{n+1}F_{m+1} + F_n F_m$
which is the statement of Identity IX for $k = n$ and proving by the strong form of induction that Identity IX holds for all positive integers n.

Section 2.3 Exercises

1. Addition table for Z_2:

	0	1
0	0	1
1	1	0

Multiplication table for Z_2:

	0	1
0	0	0
1	0	1

3. Addition table for Z_4:

	0	1	2	3
0	0	1	2	3
1	1	2	3	0
2	2	3	0	1
3	3	0	1	2

Multiplication table for Z_4:

	0	1	2	3
0	0	0	0	0
1	0	1	2	3
2	0	2	0	2
3	0	3	2	1

5. Addition table for Z_8:

	0	1	2	3	4	5	6	7
0	0	1	2	3	4	5	6	7
1	1	2	3	4	5	6	7	0
2	2	3	4	5	6	7	0	1
3	3	4	5	6	7	0	1	2
4	4	5	6	7	0	1	2	3
5	5	6	7	0	1	2	3	4
6	6	7	0	1	2	3	4	5
7	7	0	1	2	3	4	5	6

Multiplication table for Z_8:

	0	1	2	3	4	5	6	7
0	0	0	0	0	0	0	0	0
1	0	1	2	3	4	5	6	7
2	0	2	4	6	0	2	4	6
3	0	3	6	1	4	7	2	5
4	0	4	0	4	0	4	0	4
5	0	5	2	7	4	1	6	3
6	0	6	4	2	0	6	4	2
7	0	7	6	5	4	3	2	1

7. Addition table for Z_{12}:

	0	1	2	3	4	5	6	7	8	9	10	11
0	0	1	2	3	4	5	6	7	8	9	10	11
1	1	2	3	4	5	6	7	8	9	10	11	0
2	2	3	4	5	6	7	8	9	10	11	0	1
3	3	4	5	6	7	8	9	10	11	0	1	2
4	4	5	6	7	8	9	10	11	0	1	2	3
5	5	6	7	8	9	10	11	0	1	2	3	4
6	6	7	8	9	10	11	0	1	2	3	4	5
7	7	8	9	10	11	0	1	2	3	4	5	6
8	8	9	10	11	0	1	2	3	4	5	6	7
9	9	10	11	0	1	2	3	4	5	6	7	8
10	10	11	0	1	2	3	4	5	6	7	8	9
11	11	0	1	2	3	4	5	6	7	8	9	10

Multiplication table for Z_{12}:

	0	1	2	3	4	5	6	7	8	9	10	11
0	0	0	0	0	0	0	0	0	0	0	0	0
1	0	1	2	3	4	5	6	7	8	9	10	11
2	0	2	4	6	8	10	0	2	4	6	8	10
3	0	3	6	9	0	3	6	9	0	3	6	9
4	0	4	8	0	4	8	0	4	8	0	4	8
5	0	5	10	3	8	1	6	11	4	9	2	7
6	0	6	0	6	0	6	0	6	0	6	0	6
7	0	7	2	9	4	11	6	1	8	3	10	5
8	0	8	4	0	8	4	0	8	4	0	8	4
9	0	9	6	3	0	9	6	3	0	9	6	3
10	0	10	8	6	4	2	0	10	8	6	4	2
11	0	11	10	9	8	7	6	5	4	3	2	1

9. $v_1 + v_3 = (1,1,1,1,1,1,1,1,1,1,1) + (0,1,1,0,1,1,0,1,0,0,0,1) = (1,2,2,1,2,2,1,2,1,1,2)$

11. $v_3 + v_5 = (0,1,1,0,1,1,0,1,0,0,0,1) + (0,0,1,1,0,1,0,0,0,1,1,1) = (0,1,2,1,1,2,0,1,0,1,1,2)$

13. $2v_4 + v_5 + 2v_6 = 2(0,1,0,1,1,0,1,0,0,0,1,1) + (0,0,1,1,0,1,0,0,0,1,1,1) + 2(0,1,1,0,1,0,0,0,1,1,1,0)$
    $= (0,1,0,0,1,1,2,0,2,0,2,0)$

15. ATGAGGACCCTCTCTCTGCTCACTCTGCTGGCCCTG
    TACTCCTGGGAGAGAGACGAGTGAGACGACCGGGAC
    201211233303030301303230301301133301
    023033011121212123121012123123311123

17. In Z_7, F_n and L_n have period 16. A period of each is shown:
    F_n: 0, 1, 1, 2, 3, 5, 1, 6, 0, 6, 6, 5, 4, 2, 6, 1
    L_n: 2, 1, 3, 4, 0, 4, 4, 1, 5, 6, 4, 3, 0, 3, 3, 6

19. In Z_9, F_n and L_n have period 24. A period of each is shown:
    F_n: 0, 1, 1, 2, 3, 5, 8, 4, 3, 7, 1, 8, 0, 8, 8, 7, 6, 4, 1, 5, 6, 2, 8, 1
    L_n: 2, 1, 3, 4, 7, 2, 0, 2, 2, 4, 6, 1, 7, 8, 6, 5, 2, 7, 0, 7, 7, 5, 3, 8

21. The units of Z_{24} are: [1], [5], [7], [11], [13], [17], [19], [23]. Each is its own inverse.

23. In Z_{10}, $a_n = 0, 5, 5, 0, 5, 5, \ldots$ The 1s digits of the sequence a_n have period 3.

25. The 1s digits of the sequence a_n have period 12. A period of the 1s digits looks like
1347 1897 6392.

27. The 1s digits of the sequence a_n have period 60. A period of the 1s digits looks like
1785 3819 0998 7527 9651 6730 3369 5493 2572 9101 1235 8314 5943 7077 4156.

29. Each sequence
    178538190998752796516730336954932572910112358314594370774156
    932572910112358314594370774156178538190998752796516730336954
can be cyclically permuted into the other by moving the last half (30 digits) of the sequence to its beginning.

31. The period of the last two digits of the Lucas sequence is 60 as illustrated below.

	2	1	3	4	7	11	18	29	47	76	23	99	22	21	43
	64	7	71	78	49	27	76	3	79	82	61	43	4	47	51
	98	49	47	96	43	39	82	21	3	24	27	51	78	29	7
	36	43	79	22	1	23	24	47	71	18	89	7	96	3	99
	2	1	...												

Section 2.3 Advanced Exercises

1. Here $N_1 = 15$, $N_2 = 10$, $N_3 = 6$ so $[A_1] = [N_1]^{-1} = [15]^{-1} = [1]$ in Z_2, $[A_2] = [N_2]^{-1} = [10]^{-1} = [1]$ in Z_3, and $[A_3] = [N_3]^{-1} = [6]^{-1} = [1]$ in Z_1. It follows that
$N_1 A_1 a_1 + N_2 A_2 a_2 + N_3 A_3 a_3 = 15 \cdot 1 \cdot 0 + 10 \cdot 1 \cdot 1 + 6 \cdot 1 \cdot 4 = 34 \equiv 4 \pmod{30}$. The solution to the given system is $X \equiv 4 \pmod{30}$.

3. Here $N_1 = 385$, $N_2 = 231$, $N_3 = 165$, $N_4 = 105$ so $[A_1] = [N_1]^{-1} = [385]^{-1} = [1]$ in Z_3, $[A_2] = [N_2]^{-1} = [231]^{-1} = [1]$ in Z_5, $[A_3] = [N_3]^{-1} = [165]^{-1} = [2]$ in Z_7 and $[A_4] = [N_4]^{-1} = [105]^{-1} = [2]$ in Z_{11}. It follows that
$N_1 A_1 a_1 + N_2 A_2 a_2 + N_3 A_3 a_3 + N_4 A_4 a_4 = 385 \cdot 1 \cdot 2 + 231 \cdot 1 \cdot 3 + 165 \cdot 2 \cdot 4 + 105 \cdot 2 \cdot 5 = 3833 \equiv 368 \pmod{1155}$. The solution to the given system is $X \equiv 368 \pmod{1155}$.

Section 2.4 Supplementary Exercises

1. g.c.d.$(100, 13) = 1$ in 3 steps. Upper bound given by Lamé's Theorem is
   $5 \cdot$ (number of digits in 13) $= 10$ steps.

3. g.c.d.$(1220, 754) = 2$ in 12 steps. Upper bound given by Lamé's Theorem is
   $5 \cdot$ (number of digits in 754) $= 15$ steps.

5. g.c.d.$(35422, 21892) = 2$ in 19 steps. Upper bound given Lamé's Theorem is
   $5 \cdot$ (number of digits in 21892) $= 25$ steps.

7. We seek integers x and y such that $y^2 = x^2 - 37979$. Beginning with $x = 195 \geq \sqrt{37979} \approx 194.882$ and incrementing x by 1 until $y = \sqrt{x^2 - 37979}$ is an integer we find $x = 198$ and $y = 35$. Thus $37979 = (x-y)(x+y) = (163) \cdot (233)$.

9. We seek integers x and y such that $y^2 = x^2 - 4284347$. Beginning with $x = 2070 \geq \sqrt{4284347} \approx 2069.866$ and incrementing x by 1 until $y = \sqrt{x^2 - 4284347}$ is an integer we find $x = 2094$ and $y = 317$. Thus $4284347 = (x-y)(x+y) = (1777) \cdot (2411)$.

11.

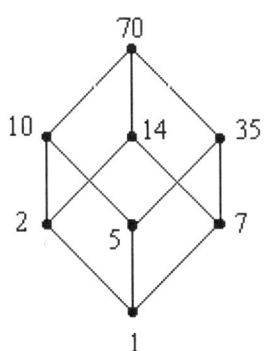

13.

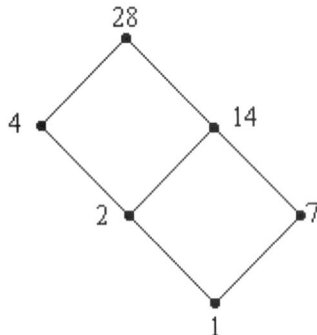

15.

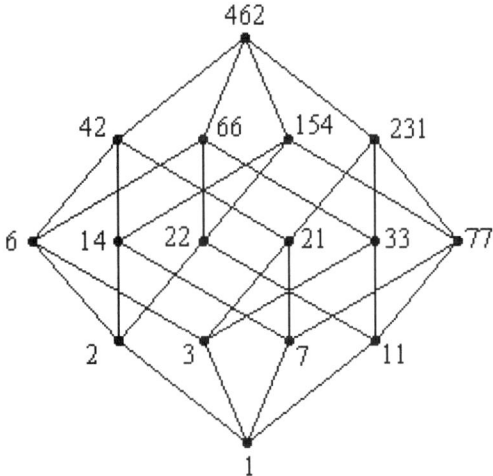

Section 2.4 Advanced Exercises

1. The value at any particular location in the triangle is found by adding the entry directly above the location with the entries diagonally to the right and above and diagonally to the left and above the location, if any exist. More generally, an entry is equal to the sum of the three closest values in the row above the entry.

For $n=0$, we have $(x^2+x+1)^n=1$ which is the first row of our triangle. This begins the induction. Fix n and consider the polynomial $(x^2+x+1)^n$, a polynomial of degree $2n$. Let b_k be the coefficient of x^k in the expansion of $(x^2+x+1)^n$. Assume the nth row of the triangle consists of the numbers b_0, b_1, ..., b_{2k}. Next consider the expansion $(x^2+x+1)^{n+1}=(b_0+b_1x+b_2x^2+...+b_{2n}x^{2n})(1+x+x^2)$. The term c_kx^k in the expansion is determined by the products $(b_kx^k)(1)$, $(b_{k-1}x^{k-1})(x)$ and $(b_{k-2}x^{k-2})(x^2)$. i.e. $c_k=b_k+b_{k-1}+b_{k-2}$. Note however, that the kth entry in the $(n+1)^{st}$ row of the triangle is also $b_k+b_{k-1}+b_{k-2}$. Thus the $(n+1)^{st}$ row consists of the coefficients in the expansion of $(x^2+x+1)^{n+1}$ completing the induction.

Chapter 3 Functions, Relations, and Counting

Section 3.1 Exercises

1. Injective: $f(a_1) = f(a_2) \Rightarrow 2a_1 + 3 = 2a_2 + 3 \Rightarrow 2a_1 = 2a_2 \Rightarrow a_1 = a_2$.
   Surjective: Let $b \in R$. Then $2a + 3 = b \Rightarrow a = \dfrac{b-3}{2} \in R$ & $f(a) = b$.
   Inverse: $f^{-1}(y) = \dfrac{y-3}{2}$.

3. Injective: $h(a_1) = h(a_2) \Rightarrow 3a_1 - 1 = 3a_2 - 1 \Rightarrow 3a_1 = 3a_2 \Rightarrow a_1 = a_2$.
   Surjective: Let $b \in R$. Then $3a - 1 = b \Rightarrow a = \dfrac{b+1}{3} \in R$ & $h(a) = b$.
   Inverse: $h^{-1}(y) = \dfrac{y+1}{3}$.

5. Injective: $l(a_1) = l(a_2) \Rightarrow 23a_1 - 13 = 23a_2 - 13 \Rightarrow 23a_1 = 23a_2 \Rightarrow a_1 = a_2$.
   Surjective: Let $b \in R$. Then $23a - 13 = b \Rightarrow a = \dfrac{b+13}{23} \in R$ & $l(a) = b$.
   Inverse: $l^{-1}(y) = \dfrac{y+13}{23}$.

7. $(f \circ g)(x) = f(x+2) = (x+2)^2 = x^2 + 4x + 4$
   $(g \circ f)(x) = g(x^2) = x^2 + 2$

   $f \circ g$ and $g \circ f$ are not equal.

9. $(f \circ g)(x) = f(x+1) = (x+1)^3 = x^3 + 3x^2 + 3x + 1$
   $(g \circ f)(x) = g(x^3) = x^3 + 1$

   $f \circ g$ and $g \circ f$ are not equal.

11.

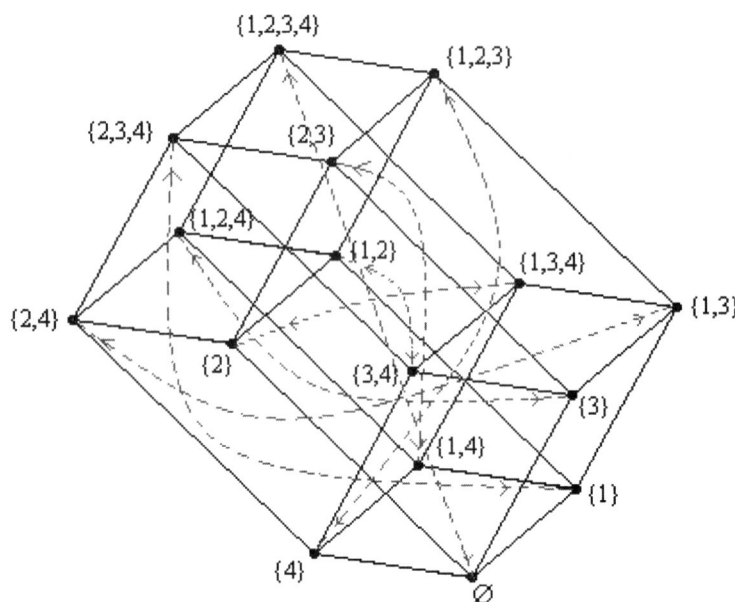

13. The subsets of X are listed to the left. To the right of each subset A is the binary vector c(A).

{1,2,3,4}	(1,1,1,1)	{2,3}	(0,1,1,0)
{1,2,3}	(1,1,1,0)	{2,4}	(0,1,0,1)
{1,2,4}	(1,1,0,1)	{3,4}	(0,0,1,1)
{1,3,4}	(1,0,1,1)	{1}	(1,0,0,0)
{2,3,4}	(0,1,1,1)	{2}	(0,1,0,0)
{1,2}	(1,1,0,0)	{3}	(0,0,1,0)
{1,3}	(1,0,1,0)	{4}	(0,0,0,1)
{1,4}	(1,0,0,1)	∅	(0,0,0,0)

15. Reflexive: no (e.g. $(4,4) \notin R$). Symmetric: no (e.g. $(1,2) \in R, (2,1) \notin R$). Antisymmetric: yes. Transitive: no (e.g. $(1,2),(2,4) \in R, (2,4) \notin R$). Partial ordering: no. Equivalence relation: no.

17. Reflexive: yes. Symmetric: yes. Antisymmetric: no (e.g. $(1,2) \in R, (2,1) \in R$). Transitive: yes. Partial ordering: no. Equivalence relation: yes.

19. R = {(1,1),(2,2),(1,2),(2,1),(3,3),(4,4),(3,4),(4,3)}

21. R = {(1,1),(2,2),(3,3),(1,2),(2,1),(1,3),(3,1),(2,3),(3,2),(4,4),(5,5),(4,5),(5,4),(6,6)}

23. (2, 3, 3)   25. (3, 3, 4, 4)

27.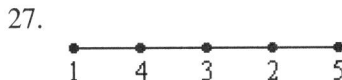

Section 3.1 Advanced Exercises

1.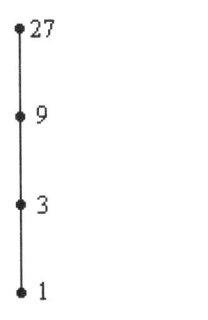

3.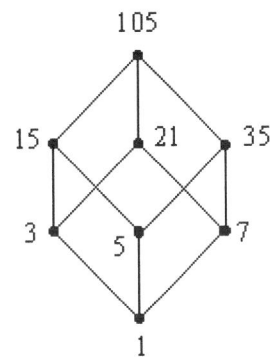

5.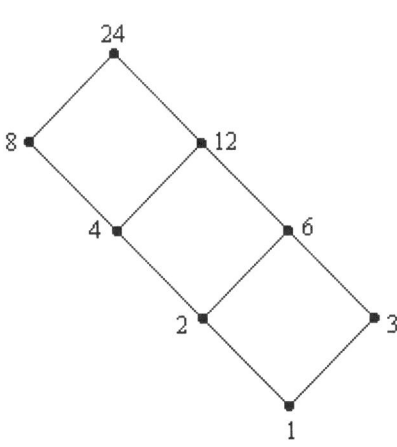

7.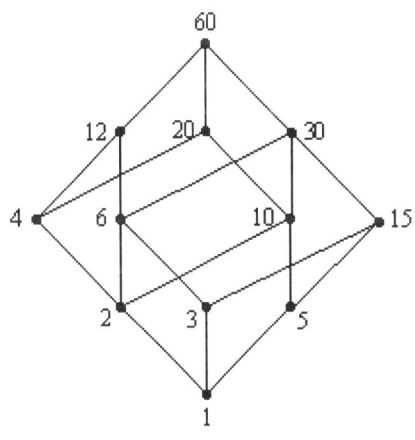

Section 3.2 Exercises

1. There are $\binom{5}{2} = 10$ possible two man rowing teams. By the Pigeonhole Principle, any assignment of the 11 meets to the 10 teams requires more than one meet to be assigned to the same team.

34 Chapter 3 Functions, Relations, and Counting

3. There are $\binom{20}{3} = 1140$ ways to choose committee members from Department A, are $\binom{17}{2} = 136$ ways to choose committee members from Department B, and $\binom{23}{4} = 8855$ ways to choose committee members from Department C. By the applied form of the Product Rule, there are $1140 \cdot 136 \cdot 8855 = 1,372,879,200$ possible committees.

5. A palindromic sequence of six letters is completely determined by selecting the first 3 letters since the last 3 letters must be the first 3 in reverse order. Further, there are no restrictions on the choices for the first three letters. Thus there are $4^3 = 64$ palindromic sequences of length six using 4 letters.

7. The set of 36 possible outcomes for the roll of two dice is the set $A \times A$ where $A = \{1,2,3,4,5,6\}$. Let $O = \{1,3,5\}$ and $E = \{2,4,6\}$. The set $(O \times O) \cup (E \times E)$ containing $3 \cdot 3 + 3 \cdot 3 = 18$ elements is the set of all even rolls of the two die. Since ½ the rolls are even, ½ are odd and the game is fair.

9. $P(6,3) = \dfrac{6!}{3!} = 120$

11. $P(7,4) = \dfrac{7!}{3!} = 840$

13. $P(10,3) = \dfrac{10!}{7!} = 720$

15. $P(14,6) = \dfrac{14!}{8!} = 2,162,160$

17. $P(20,10) = \dfrac{20!}{10!} = 670,442,572,800$

19. Here the order in which the cards are dealt does not matter. The number of 13 card hands is $C(52,13) = 635,013,559,600$.

21. $\dfrac{20!}{10!5!5!} = 46,558,512$

23. The numbers of spellings is the number of paths through a 4 by 4 grid which is $\binom{4+4}{4} = 70$.

Section 3.2 Advanced Exercises

1. There are $\binom{3+3-1}{3} = 10$ 3-element multisets taken from $\{1,2,3\}$. They are

{1,1,1}   {2,2,2}   {3,3,3}   {1,1,2}   {1,1,3}
{1,2,2}   {2,2,3}   {1,3,3}   {2,3,3}   {1,2,3}

3. . There are $\binom{4+4-1}{4} = 35$ 4-element multisets taken from $\{1,2,3,4\}$. They are

$\{1,1,1,1\}$	$\{2,2,2,2\}$	$\{3,3,3,3\}$	$\{4,4,4,4\}$	$\{1,1,1,2\}$	$\{1,1,1,3\}$	$\{1,1,1,4\}$
$\{1,2,2,2\}$	$\{2,2,2,3\}$	$\{2,2,2,4\}$	$\{1,3,3,3\}$	$\{2,3,3,3\}$	$\{3,3,3,4\}$	$\{1,4,4,4\}$
$\{2,4,4,4\}$	$\{3,4,4,4\}$	$\{1,1,2,2\}$	$\{1,1,3,3\}$	$\{1,1,4,4\}$	$\{2,2,3,3\}$	$\{2,2,4,4\}$
$\{3,3,4,4\}$	$\{1,1,2,3\}$	$\{1,1,2,4\}$	$\{1,1,3,4\}$	$\{1,2,2,3\}$	$\{1,2,2,4\}$	$\{2,2,3,4\}$
$\{1,2,3,3\}$	$\{1,3,3,4\}$	$\{2,3,3,4\}$	$\{1,2,4,4\}$	$\{1,3,4,4\}$	$\{2,3,4,4\}$	$\{1,2,3,4\}$

Section 3.3 Exercises

1. $\sum_{k=0}^{5}\binom{5}{k}^2 = \binom{5}{0}^2 + \binom{5}{1}^2 + \binom{5}{2}^2 + \binom{5}{3}^2 + \binom{5}{4}^2 + \binom{5}{5}^2 = 1+25+100+100+25+1$
$= 252 = \binom{10}{5}$

3. $\sum_{k=0}^{7}\binom{7}{k}^2 = \binom{7}{0}^2 + \binom{7}{1}^2 + \binom{7}{2}^2 + \binom{7}{3}^2 + \binom{7}{4}^2 + \binom{7}{5}^2 + \binom{7}{6}^2 + \binom{7}{7}^2$
$= 1+49+441+1225+1225+441+49+1 = 3432 = \binom{14}{7}$

5. $\sum_{k=0}^{4}\binom{2+k}{k} = \binom{2}{0} + \binom{3}{1} + \binom{4}{2} + \binom{5}{3} + \binom{6}{4} = 1+3+6+10+15 = 35 = \binom{7}{4}$

7. $\sum_{k=0}^{3}(-1)^k\binom{3}{k}\frac{6}{k+3} = 2\binom{3}{0} - \frac{3}{2}\binom{3}{1} + \frac{6}{5}\binom{3}{2} - \binom{3}{3} = 2 - \frac{9}{2} + \frac{18}{5} - 1 = \frac{1}{10} = \binom{5}{2}^{-1}$

9.

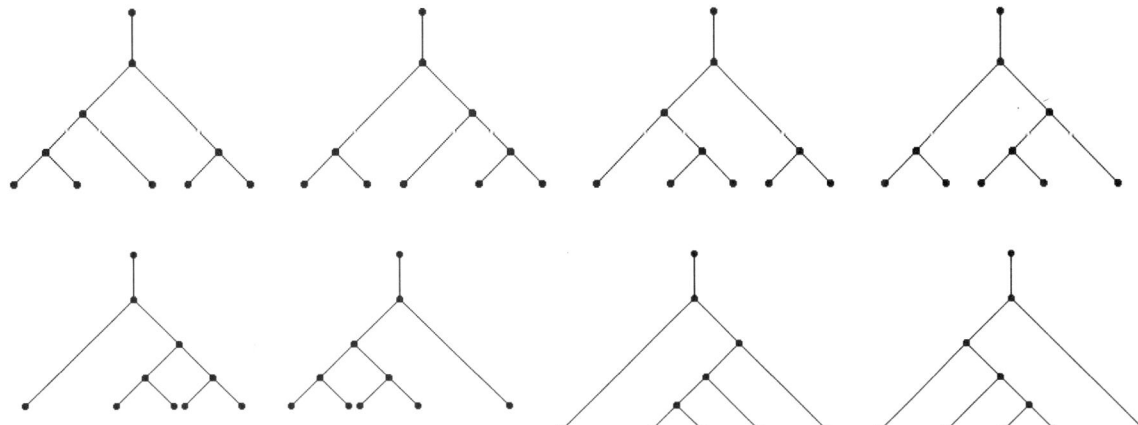

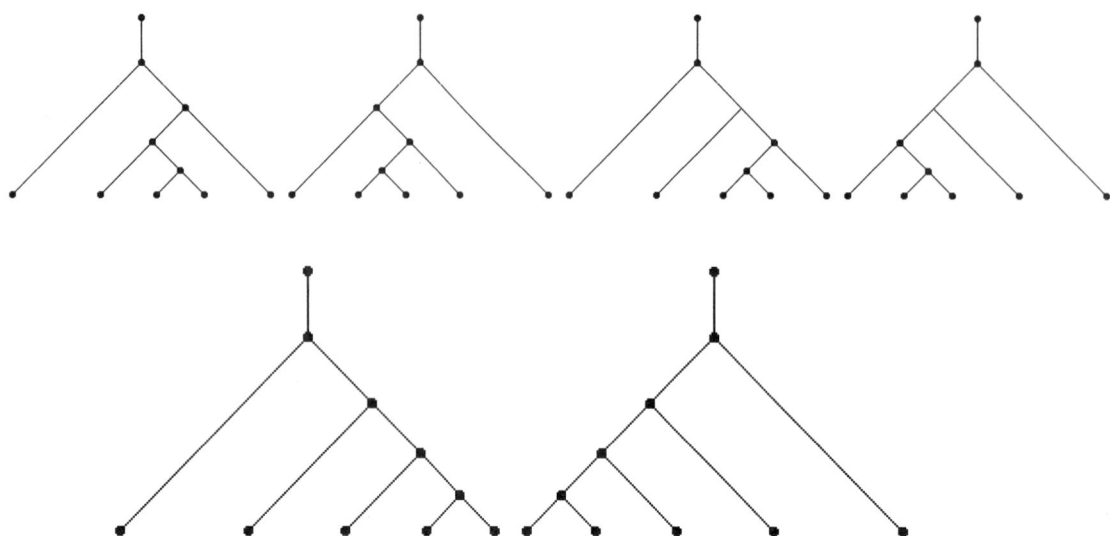

11. $C_7 = \dfrac{1}{8}\dbinom{14}{7} = \dfrac{14!}{8!7!} = 429$

$C_7 = \dbinom{14}{7} - \dbinom{14}{6} = 3432 - 3003 = 429$

13. $C_6 = C_0C_5 + C_1C_4 + C_2C_3 + C_3C_2 + C_4C_1 + C_5C_0 = 1\cdot 42 + 1\cdot 14 + 2\cdot 5 + 5\cdot 2 + 14\cdot 1 + 42\cdot 1 = 132$

15. $C_8 = C_0C_7 + C_1C_6 + C_2C_5 + C_3C_4 + C_4C_3 + C_5C_2 + C_6C_1 + C_7C_0$
$= 1\cdot 429 + 1\cdot 132 + 2\cdot 42 + 5\cdot 14 + 14\cdot 5 + 42\cdot 2 + 132\cdot 1 + 429\cdot 1 = 1430$

17. Since $6 < \sqrt{47} < 7$ we focus on the primes 2, 3, and 5. Let A_1, A_2 and A_3 denote the set of positive integers ≤ 47 that are divisible by 2, 3 and 5 respectively. Then $|A_1| = \lfloor 47/2 \rfloor = 23$, $|A_2| = \lfloor 47/3 \rfloor = 15$, $|A_3| = \lfloor 47/5 \rfloor = 9$, $|A_1 \cap A_2| = \lfloor 47/(2\cdot 3) \rfloor = 7$, $|A_1 \cap A_3| = \lfloor 47/(2\cdot 5) \rfloor = 4$, $|A_2 \cap A_3| = \lfloor 47/(3\cdot 5) \rfloor = 3$ and $|A_1 \cap A_2 \cap A_3| = \lfloor 47/(2\cdot 3\cdot 5) \rfloor = 1$ By the Principle of Inclusion- Exclusion, the number of integers ≤ 47 divisible by any of 2, 3 and 5 is $|A_1 \cup A_2 \cup A_3| = 23 + 15 + 9 - 7 - 4 - 3 + 1 = 34$. The number of primes ≤ 47, taking into account the primes 2, 3 and 5 and ignoring 1 is then $47 - 34 + 3 - 1 = 15$.

19. Since $9 < \sqrt{82} < 10$ we focus on the primes 2, 3, 5 and 7. Let A_1, A_2, A_3 and A_4 denote the set of positive integers ≤ 82 that are divisible by 2, 3, 5 and 7 respectively. Then $|A_1| = \lfloor 82/2 \rfloor = 41$, $|A_2| = \lfloor 82/3 \rfloor = 27$, $|A_3| = \lfloor 82/5 \rfloor = 16$, $|A_4| = \lfloor 82/7 \rfloor = 11$, $|A_1 \cap A_2| = \lfloor 82/(2\cdot 3) \rfloor = 13$, $|A_1 \cap A_3| = \lfloor 82/(2\cdot 5) \rfloor = 8$, $|A_1 \cap A_4| = \lfloor 82/(2\cdot 7) \rfloor = 5$, $|A_2 \cap A_3| = \lfloor 82/(3\cdot 5) \rfloor = 5$, $|A_2 \cap A_4| = \lfloor 82/(3\cdot 7) \rfloor = 3$, $|A_3 \cap A_4| = \lfloor 82/(5\cdot 7) \rfloor = 2$, $|A_1 \cap A_2 \cap A_3| = \lfloor 82/(2\cdot 3\cdot 5) \rfloor = 2$, $|A_1 \cap A_2 \cap A_4| = \lfloor 82/(2\cdot 3\cdot 7) \rfloor = 1$, $|A_1 \cap A_3 \cap A_4| = \lfloor 82/(2\cdot 5\cdot 7) \rfloor = 1$, $|A_2 \cap A_3 \cap A_4| = \lfloor 82/(3\cdot 5\cdot 7) \rfloor = 0$ and $|A_1 \cap A_2 \cap A_3 \cap A_4| = \lfloor 82/(2\cdot 3\cdot 5\cdot 7) \rfloor = 0$.

By the Principle of Inclusion- Exclusion, the number of integers ≤ 82 divisible by any of 2, 3, 5 and 7 is
$|A_1 \cup A_2 \cup A_3 \cup A_4| = 41 + 27 + 16 + 11 - 13 - 8 - 5 - 5 - 3 - 2 + 2 + 1 + 1 + 0 - 0 = 63$. The number of primes ≤ 62, taking into account the primes 2, 3, 5 and 7 and ignoring 1 is then $82 - 63 + 4 - 1 = 22$.

21. Using the values of the Bernoulli numbers supplied in the text we have
$B_4(X) = \sum_{k=0}^{4} B_k \binom{4}{k} X^{4-k} = X^4 - 2X^3 + X^2 - \frac{1}{30}$. By Bernoulli's formula
$P_3(N) = \frac{1}{4}\left(B_4(N+1) - \left(-\frac{1}{30}\right)\right) = \frac{(N+1)^4 - 2(N+1)^3 + (N+1)^2}{4} = \left(\frac{N(N+1)}{2}\right)^2$.

23. Using the values of the Bernoulli numbers supplied in the text we have
$B_6(X) = \sum_{k=0}^{6} B_k \binom{6}{k} X^{6-k} = X^6 - 3X^5 + \frac{5}{2}X^4 - \frac{1}{2}X^2 + \frac{1}{42}$. By Bernoulli's formula
$P_5(100) = \frac{1}{6}\left(B_6(101) - \frac{1}{42}\right) = 171,708,332,500$.

Section 3.3 Advanced Exercises

1. $\binom{n}{k} = \frac{n!}{k!(n-k)!} = \frac{n(n-1)!}{k(k-1)!(n-k)!} = \frac{n}{k}\frac{(n-1)!}{(k-1)!(n-k)!} = \frac{n}{k}\binom{n-1}{k-1}$

3. We use induction on n to show $\sum_{k=0}^{n} \binom{r+k}{k} = \binom{r+n+1}{n}$. The identity is true for $n = 0$ since
$\sum_{k=0}^{0} \binom{r+k}{k} = \binom{r}{0} = 1 = \binom{r+1}{0}$. Now assume $\sum_{k=0}^{n} \binom{r+k}{k} = \binom{r+n+1}{n}$ for fixed n. Then
$\sum_{k=0}^{n+1} \binom{r+k}{k} = \sum_{k=0}^{n} \binom{r+k}{k} + \binom{r+n+1}{n+1} = \binom{r+n+1}{n} + \binom{r+n+1}{n+1}$ by the induction hypothesis. Recalling the basic identity $\binom{n}{k} = \binom{n-1}{k-1} + \binom{n-1}{k}$, the right hand side above can be written as $\binom{r+n+2}{n+1}$ which is the right hand side of the Parallel Summation formula with n replaced by $n+1$ completing the induction argument.

Section 3.4 Supplementary Exercises

1. Injective: $f(a_1) = f(a_2) \Rightarrow 2a_1 + 4 = 2a_2 + 4 \Rightarrow 2a_1 = 2a_2 \Rightarrow a_1 = a_2$.
   Surjective: Let $b \in R$. Then $2a + 4 = b \Rightarrow a = \frac{b-4}{2} \in R$ & $f(a) = b$.
   Inverse: $f^{-1}(y) = \frac{y-4}{2}$.

3. Injective: $h(a_1) = h(a_2) \Rightarrow 8a_1 + 16 = 8a_2 + 16 \Rightarrow 8a_1 = 8a_2 \Rightarrow a_1 = a_2$.

Surjective: Let $b \in R$. Then $8a + 16 = b \Rightarrow a = \dfrac{b-16}{8} \in R$ & $h(a) = b$.

Inverse: $h^{-1}(y) = \dfrac{y-16}{8}$.

5.

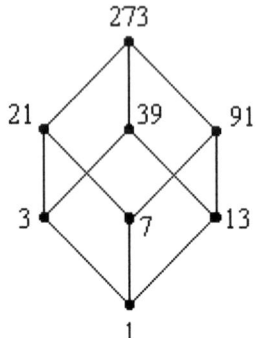

7.

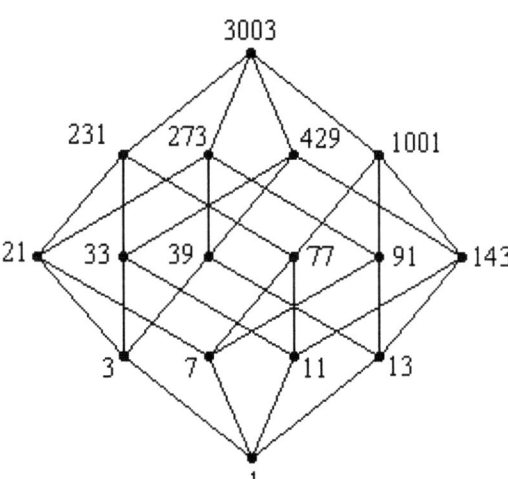

9.

11. (6, 6, 6, 6)

13. There are fifteen ways to roll a sum of eight or larger represented by the set
{(2,6), (3,5), (4,4), (5,3), (6,2), (3,6), (4,5), (5,4), (6,3), (4,6), (5,5), (6,4), (5,6), (6,5), (6,6)}.

15. There are ten ways to roll a sum of five or less represented by the set
{(1,1), (1,2), (2,1), (1,3), (2,2), (3,1), (1,4), (2,3), (3,2), (4,1)}.

17. There are 35 rolls of three dice with a sum of seven or less. The list of such rolls is given in the table below.

(1,1,1)	(1,1,2)	(1,2,1)	(2,1,1)	(1,2,2)	(2,1,2)	(2,2,1)
(1,1,3)	(1,3,1)	(3,1,1)	(1,1,4)	(1,4,1)	(4,1,1)	(1,2,3)
(1,3,2)	(2,1,3)	(2,3,1)	(3,1,2)	(3,2,1)	(2,2,2)	(1,1,5)
(1,5,1)	(5,1,1)	(1,3,3)	(3,1,3)	(3,3,1)	(2,2,3)	(2,3,2)
(3,2,2)	(1,2,4)	(1,4,2)	(2,1,4)	(2,4,1)	(4,1,2)	(4,2,1)

19. There are $\dfrac{5!}{2!2!1!} = 30$ such partitions. They are

({1,2},{3,4},{5})	({3,4},{1,2},{5})	({1,3},{2,4},{5})	({2,4},{1,3},{5})	({1,4},{2,3},{5})
({2,3},{1,4},{5})	({1,2},{3,5},{4})	({3,5},{1,2},{4})	({1,3},{2,5},{4})	({2,5},{1,3},{4})
({1,5},{2,3},{4})	({2,3},{1,5},{4})	({1,2},{4,5},{3})	({4,5},{1,2},{3})	({1,4},{2,5},{3})
({2,5},{1,4},{3})	({1,5},{2,4},{3})	({2,4},{1,5},{3})	({1,3},{4,5},{2})	({4,5},{1,3},{2})
({1,4},{3,5},{2})	({3,5},{1,4},{2})	({1,5},{3,4},{2})	({3,4},{1,5},{2})	({2,3},{4,5},{1})
({4,5},{2,3},{1})	({2,4},{3,5},{1})	({3,5},{2,4},{1})	({2,5},{3,4},{1})	({3,4},{2,5},{1})

21. There are $\binom{5+3-1}{3} = 35$ 3-element multisets taken from a set with five elements such as $\{1,2,3,4,5\}$.
They are

{1,1,1}	{2,2,2}	{3,3,3}	{4,4,4}	{5,5,5}	{1,1,2}	{1,1,3}
{1,1,4}	{1,1,5}	{1,2,2}	{2,2,3}	{2,2,4}	{2,2,5}	{1,3,3}
{2,3,3}	{3,3,4}	{3,3,5}	{1,4,4}	{2,4,4}	{3,4,4}	{4,4,5}
{1,5,5}	{2,5,5}	{3,5,5}	{4,5,5}	{1,2,3}	{1,2,4}	{1,2,5}
{1,3,4}	{1,3,5}	{1,4,5}	{2,3,4}	{2,3,5}	{2,4,5}	{3,4,5}

23. Since $11 < \sqrt{130} < 12$ we consider the primes 2, 3, 5, 7 and 11. Let A_1, A_2, A_3, A_4 and A_5 denote the set of positive integers ≤ 130 that are divisible by 2, 3, 5, 7 and 11 respectively. Then

$|A_1| = \lfloor 130/2 \rfloor = 65$, $|A_2| = \lfloor 130/3 \rfloor = 43$, $|A_3| = \lfloor 130/5 \rfloor = 26$, $|A_4| = \lfloor 130/7 \rfloor = 18$, $|A_5| = \lfloor 130/11 \rfloor = 11$, which sum to 163,

$|A_1 \cap A_2| = \lfloor 130/(2 \cdot 3) \rfloor = 21$, $|A_1 \cap A_3| = \lfloor 130/(2 \cdot 5) \rfloor = 13$, $|A_1 \cap A_4| = \lfloor 130/(2 \cdot 7) \rfloor = 9$,
$|A_1 \cap A_5| = \lfloor 130/(2 \cdot 11) \rfloor = 5$, $|A_2 \cap A_3| = \lfloor 130/(3 \cdot 5) \rfloor = 8$, $|A_2 \cap A_4| = \lfloor 130/(3 \cdot 7) \rfloor = 6$,
$|A_2 \cap A_5| = \lfloor 130/(3 \cdot 11) \rfloor = 3$, $|A_3 \cap A_4| = \lfloor 130/(5 \cdot 7) \rfloor = 3$, $|A_3 \cap A_5| = \lfloor 130/(5 \cdot 11) \rfloor = 2$,
$|A_4 \cap A_5| = \lfloor 130/(7 \cdot 11) \rfloor = 1$, which sum to 71,

$|A_1 \cap A_2 \cap A_3| = \lfloor 130/(2 \cdot 3 \cdot 5) \rfloor = 4, |A_1 \cap A_2 \cap A_4| = \lfloor 130/(2 \cdot 3 \cdot 7) \rfloor = 3, |A_1 \cap A_2 \cap A_5| = \lfloor 130/(2 \cdot 3 \cdot 11) \rfloor = 1,$
$|A_1 \cap A_3 \cap A_4| = \lfloor 130/(2 \cdot 5 \cdot 7) \rfloor = 1, |A_1 \cap A_3 \cap A_5| = \lfloor 130/(2 \cdot 5 \cdot 11) \rfloor = 1, |A_1 \cap A_4 \cap A_5| = \lfloor 130/(2 \cdot 7 \cdot 11) \rfloor = 0,$
$|A_2 \cap A_3 \cap A_4| = \lfloor 130/(3 \cdot 5 \cdot 7) \rfloor = 1, |A_2 \cap A_3 \cap A_5| = \lfloor 130/(3 \cdot 5 \cdot 11) \rfloor = 0, |A_2 \cap A_4 \cap A_5| = \lfloor 130/(3 \cdot 7 \cdot 11) \rfloor = 0$
$, |A_3 \cap A_4 \cap A_5| = \lfloor 130/(5 \cdot 7 \cdot 11) \rfloor = 0,$
which sum to 11,

$|A_1 \cap A_2 \cap A_3 \cap A_4| = \lfloor 130/(2 \cdot 3 \cdot 5 \cdot 7) \rfloor = 0, |A_1 \cap A_2 \cap A_3 \cap A_5| = \lfloor 130/(2 \cdot 3 \cdot 5 \cdot 11) \rfloor = 0,$
$|A_1 \cap A_2 \cap A_4 \cap A_5| = \lfloor 130/(2 \cdot 3 \cdot 7 \cdot 11) \rfloor = 0, |A_1 \cap A_3 \cap A_4 \cap A_5| = \lfloor 130/(2 \cdot 5 \cdot 7 \cdot 11) \rfloor = 0,$
$|A_2 \cap A_3 \cap A_4 \cap A_5| = \lfloor 130/(3 \cdot 5 \cdot 7 \cdot 11) \rfloor = 0$, which sum to 0, and

$|A_1 \cap A_2 \cap A_3 \cap A_4 \cap A_5| = \lfloor 130/(2 \cdot 3 \cdot 5 \cdot 7 \cdot 11) \rfloor = 0$.

By the Principle of Inclusion- Exclusion, the number of integers ≤ 130 divisible by any of 2, 3, 5, 7 and 11 is $|A_1 \cup A_2 \cup A_3 \cup A_4 \cup A_5| = 163 - 71 + 11 - 0 + 0 = 103$. The number of primes ≤ 130, taking into account the primes 2, 3, 5, 7 and 11 and ignoring 1 is then $130 - 103 + 5 - 1 = 31$.

25. Using the values of the Bernoulli numbers supplied in the text we have
$B_5(X) = \sum_{k=0}^{5} B_k \binom{5}{k} X^{5-k} = X^5 - \frac{5}{2}X^4 + \frac{5}{3}X^3 - \frac{1}{6}X$. By Bernoulli's formula
$P_4(N) = \frac{1}{5} B_5(N+1) = \frac{N(N+1)(2N+1)(3N^2 + 3N - 1)}{30}$.

27. Using the values of the Bernoulli numbers supplied in the text we have
$B_7(X) = \sum_{k=0}^{7} B_k \binom{7}{k} X^{7-k} = X^7 - \frac{7}{2}X^6 + \frac{7}{2}X^5 - \frac{7}{6}X^3 + \frac{1}{6}X$. By Bernoulli's formula
$P_6(100) = \frac{1}{7} B_7(101) = 14,790,714,119,050$.

Chapter 4 Graphs

Section 4.1 Exercises

Answers to exercises 1-14 may vary with respect to the labeling of nodes and the choice of directions in creation of the digraphs.

1. Edge list for graph below: (1,2), (1,3), (1,4), (1,5), (2,3), (2,4), (2,5), (3,4), (3,5), (4,5) .

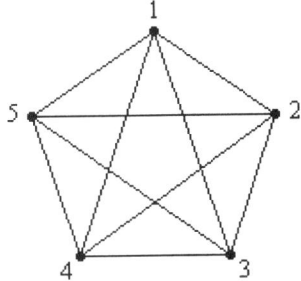

Edge list for digraph below: (1,3), (1,4), (1,5), (2,1), (2,3), (2,4), (2,5), (4,3), (5,3), (5,4) .

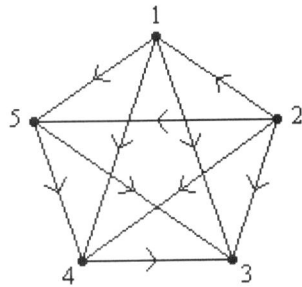

3. Edge list for graph below: (1,4), (1,5), (1,6), (2,4), (2,5), (2,6), (3,4), (3,5), (3,6) .

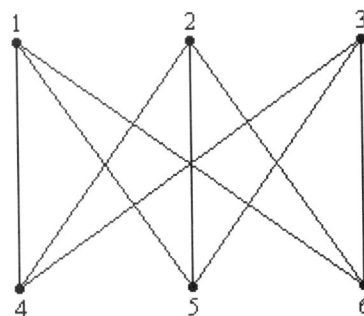

Edge list for digraph below: (1,4), (1,5), (1,6), (2,4), (2,5), (2,6), (3,4), (3,5), (3,6) .

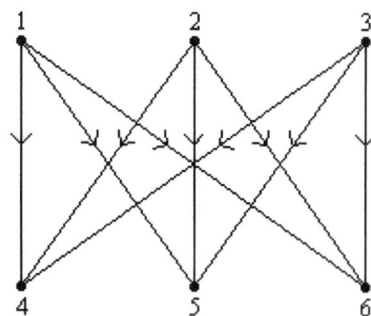

5. Edge list for graph below: (1,2), (1,3), (1,4), (1,5), (2,6), (2,7), (2,9), (3,6), (3,8), (3,10), (4,7), (4,8), (4,11), (5,9), (5,10), (5,11), (6,12), (6,13), (7,12), (7,14), (8,12), (8,15), (9,13), (9,14), (10,13), (10,15), (11,14), (11,15), (12,16), (13,16), (14,16), (15,16).

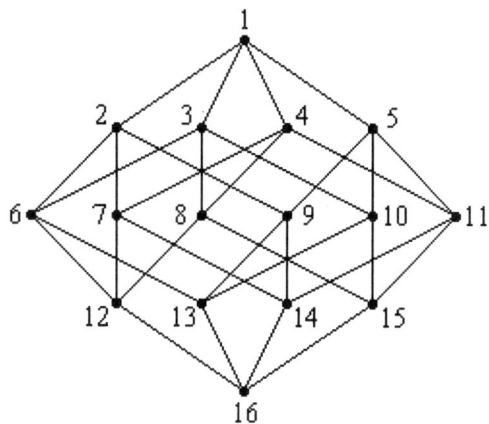

Edge list for digraph below: (1,2), (1,3), (1,4), (1,5), (2,6), (2,7), (2,9), (3,6), (3,8), (3,10), (4,7), (4,8), (4,11), (5,9), (5,10), (5,11), (6,12), (6,13), (7,12), (7,14), (8,12), (8,15), (9,13), (9,14), (10,13), (10,15), (11,14), (11,15), (12,16), (13,16), (14,16), (15,16).

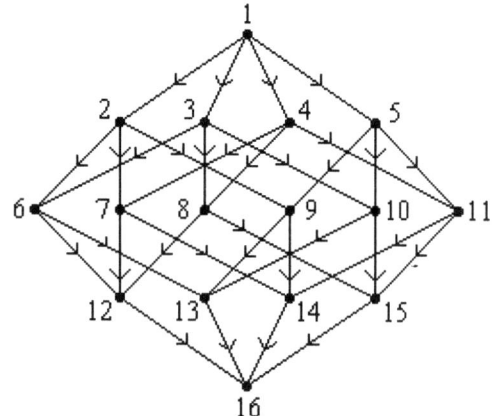

7. Edge list for the graph below: (1,2), (1,3), (1,4), (1,6), (1,7), (1,8), (2,3), (2,4), (2,5), (2,7), (2,8), (3,4), (3,5), (3,6), (3,8), (4,5), (4,6), (4,7), (5,6), (5,7), (5,8), (6,7), (6,8), (7,8).

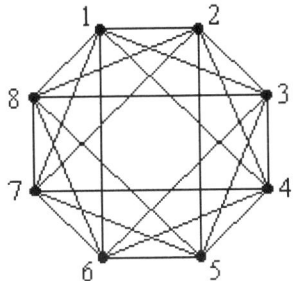

Edge list for the digraph below: (1,3), (1,4), (1,6), (1,7), (1,8), (2,1), (2,3), (2,4), (2,5), (2,7), (2,8), (4,3), (5,3), (5,4), (6,3), (6,4), (6,5), (7,4), (7,5), (7,6), (8,3), (8,5), (8,6), (8,7).

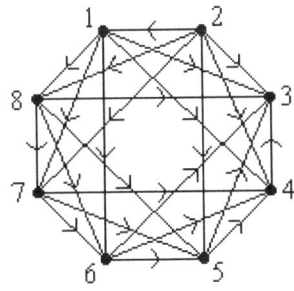

9. Undirected graph:

$$\begin{pmatrix} 0 & 1 & 1 & 1 & 1 & 1 \\ 1 & 0 & 1 & 1 & 1 & 1 \\ 1 & 1 & 0 & 1 & 1 & 1 \\ 1 & 1 & 1 & 0 & 1 & 1 \\ 1 & 1 & 1 & 1 & 0 & 1 \\ 1 & 1 & 1 & 1 & 1 & 0 \end{pmatrix}$$

Directed graph:

$$\begin{pmatrix} 0 & 0 & 1 & 1 & 1 & 1 \\ 1 & 0 & 1 & 1 & 1 & 1 \\ 0 & 0 & 0 & 0 & 0 & 0 \\ 0 & 0 & 1 & 0 & 0 & 0 \\ 0 & 0 & 1 & 1 & 0 & 0 \\ 0 & 0 & 1 & 1 & 1 & 0 \end{pmatrix}$$

11. Undirected graph:

$$\begin{pmatrix} 0 & 1 & 1 & 1 & 0 & 0 & 0 & 0 \\ 1 & 0 & 0 & 0 & 1 & 1 & 0 & 0 \\ 1 & 0 & 0 & 0 & 1 & 0 & 1 & 0 \\ 1 & 0 & 0 & 0 & 0 & 1 & 1 & 0 \\ 0 & 1 & 1 & 0 & 0 & 0 & 0 & 1 \\ 0 & 1 & 0 & 1 & 0 & 0 & 0 & 1 \\ 0 & 0 & 1 & 1 & 0 & 0 & 0 & 1 \\ 0 & 0 & 0 & 0 & 1 & 1 & 1 & 0 \end{pmatrix}$$

Directed graph:

$$\begin{pmatrix} 0 & 1 & 0 & 0 & 0 & 0 & 0 & 0 \\ 0 & 0 & 0 & 0 & 0 & 1 & 0 & 0 \\ 1 & 0 & 0 & 0 & 1 & 0 & 0 & 0 \\ 1 & 0 & 0 & 0 & 0 & 0 & 0 & 0 \\ 0 & 1 & 0 & 0 & 0 & 0 & 0 & 1 \\ 0 & 0 & 0 & 1 & 0 & 0 & 0 & 0 \\ 0 & 0 & 1 & 1 & 0 & 0 & 0 & 0 \\ 0 & 0 & 0 & 0 & 0 & 1 & 1 & 0 \end{pmatrix}$$

44   Chapter 4 Graphs

13. Undirected graph: Directed graph:

$$\begin{pmatrix} 0 & 1 & 1 & 0 & 1 & 1 \\ 1 & 0 & 1 & 1 & 0 & 1 \\ 1 & 1 & 0 & 1 & 1 & 0 \\ 0 & 1 & 1 & 0 & 1 & 1 \\ 1 & 0 & 1 & 1 & 0 & 1 \\ 1 & 1 & 0 & 1 & 1 & 0 \end{pmatrix} \qquad \begin{pmatrix} 0 & 0 & 1 & 0 & 1 & 1 \\ 1 & 0 & 1 & 1 & 0 & 1 \\ 0 & 0 & 0 & 0 & 0 & 0 \\ 0 & 0 & 1 & 0 & 0 & 0 \\ 0 & 0 & 1 & 1 & 0 & 0 \\ 0 & 0 & 0 & 1 & 1 & 0 \end{pmatrix}$$

15.

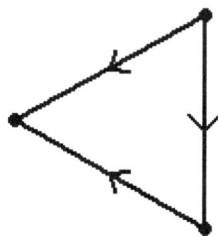

17.

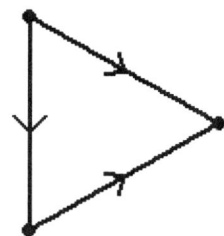

19. The sequence (1,1,3,3) can only be the degree sequence of a disconnected pseudograph such as

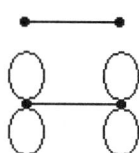

21. The sequence (1,1,1,1,1,1,1,1,1,3,3,4,5) can be realized by the tree pictured below.

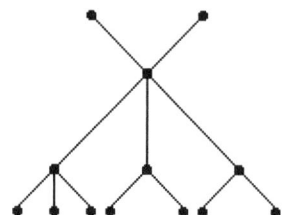

23. The graphs of the octahedron (left) and K_4 (right) are shown below with labeled nodes.

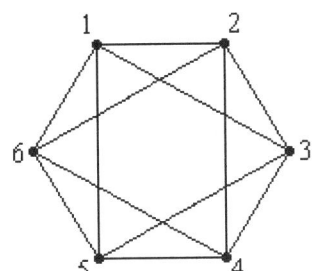 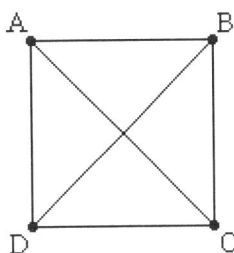

The homeomorph of K_4 in the octahedron is illustrated by specifying the mapping of K_4 nodes to nodes of the octahedron and listing the disjoint paths in the octahedron corresponding to the edges of K_4.

Nodes: A→1 B→3 C→4 D→6
Edges: (A, B)→(1, 3)   (A, C)→(1, 2)(2, 4)   (A, D)→(1, 6)
       (B, C)→(3, 4)   (B, D)→(3, 5)(5, 6)   (C, D)→(4, 6)

Section 4.1 Advanced Exercises

1. In K_n there is one edge for every pair of nodes. Thus the number of edges of K_n is $\binom{n}{2} = \frac{n(n-1)}{2}$.

3. The n-dimensional octahedron can be constructed from K_{2n}, the complete graph on $2n$ nodes by removing the n edges connecting opposing nodes. Thus, using the result of exercise 1, there are $\binom{2n}{2} - n = \frac{2n(2n-1)}{2} - n = 2n(n-1)$ nodes in the n-dimensional octahedron.

5. Any simple graph on 4 nodes is isomorphic to one of the following graphs:

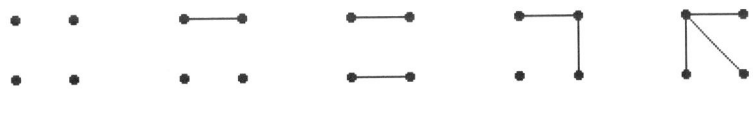

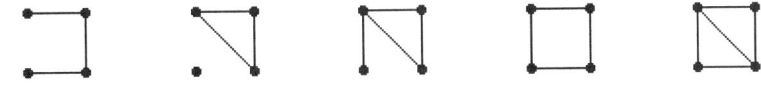

Section 4.2 Exercises

1. The graph on the left below shows the given graph with one additional bridge while the version on the left contains two additional bridges. In the first, an Eulerian circuit is indicated that visits the nodes 1, 3, 2, 3, 4, 2, 1, 4 in order. In the second, another bridge is added from 4 to 1 extending the circuit to an Eulerian cycle.

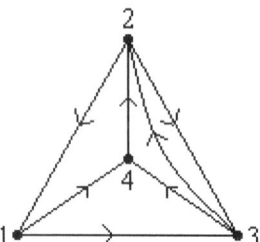 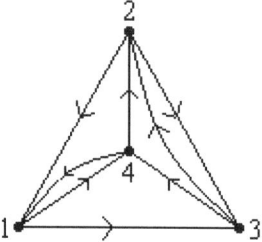

3. The Eulerian cycle in K_5

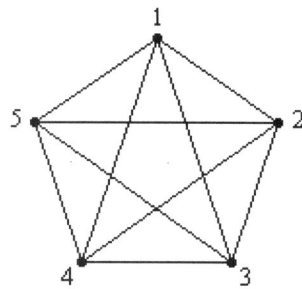

can be described by the sequence of nodes: 1, 2, 3, 4, 5, 1, 3, 5, 2, 6, 1.

5. Given the labeling of the 4-dimensional octahedron shown below

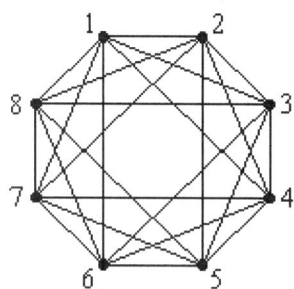

an Eulerian cycle is defined by visiting the nodes in the order: 1, 4, 5, 8, 3, 4, 7, 2, 3, 6, 1, 2, 5, 3, 1, 8, 6, 7, 5, 6, 4, 2, 8, 7, 1.

7. Assume the 5-cube has been constructed with the nodes labeled with binary 5-vectors. The Standard Grey Code provides a sequence of nodes which define a Hamiltonian circuit: (0,0,0,0,0), (0,0,0,0,1), (0,0,0,1,1), (0,0,0,1,0), (0,0,1,1,0), (0,0,1,1,1), (0,0,1,0,1), (0,0,1,0,0), (0,1,1,0,0), (0,1,1,0,1), (0,1,1,1,1), (0,1,1,1,0), (0,1,0,1,0), (0,1,0,1,1), (0,1,0,0,1), (0,1,0,0,0), (1,1,0,0,0), (1,1,0,0,1), (1,1,0,1,1), (1,1,0,1,0), (1,1,1,1,0), (1,1,1,1,1), (1,1,1,0,1), (1,1,1,0,0), (1,0,1,0,0), (1,0,1,0,1), (1,0,1,1,1), (1,0,1,1,0), (1,0,0,1,0), (1,0,0,1,1), (1,0,0,0,1), (1,0,0,0,0)
which becomes a cycle by continuing to (0,0,0,0,0).

9.

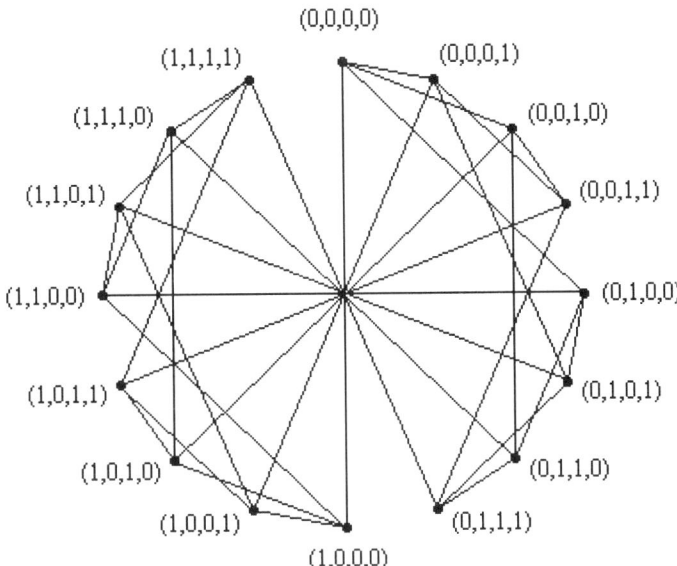

11. Number the street corners 1-12 from north to south and west to east in order. A solution to the Two-Way Street Problem is given by the following trail of nodes: 5, 1, 2, 3, 4, 8, 12, 11, 10, 9, 5, 6, 7, 8, 7, 6, 5, 9, 10, 11, 12, 8, 4, 3, 2, 1, 5.

13. The Travelling Salesman Problem is solved for this graph by traversing the nodes around the outer edge, this is use the edges labeled 15, 20, 14 and 18.

15. Assume the 4-dimensional octahedron is planar. We have $V = 8$ vertices and $E = 24$ edges. The number of faces is then $F = E + 2 - V = 18$. Any cycle in any graph has at least 3 edges, so we must have $3F \leq 2E$. Note however, for the 4-dimensional octahedron, $3F = 3(18) = 54 > 48 = 2E$, a contradiction, so the 4-dimensional octahedron is not planar.

48 Chapter 4 Graphs

Section 4.2 Advanced Exercises

1. The nodes of the dodecahedron can be viewed as four nested pentagons (Figure 4.16 in the text). A Hamiltonian circuit is pictured below.

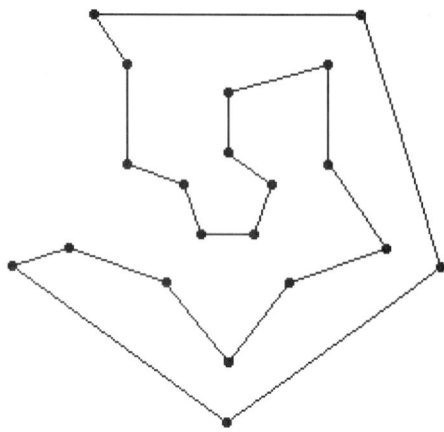

3. The problem gives rise to the graph

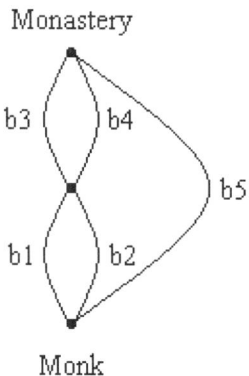

where each edge is labeled for a bridge. One path the monk may take to cross every bridge and reach the monastery is b1, b3, b5, b2, b4.

5. The corresponding labeled graph for this problem is

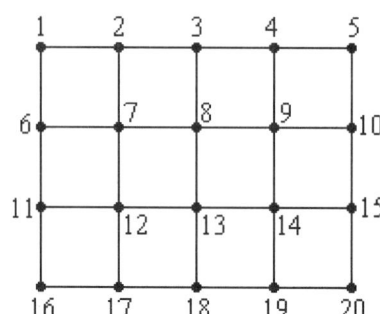

and the cycle starting at 1 with minimal retracing is specified by the sequence of nodes: 1, 2, 3, 4, 9, 14, 19, 18, 13, 8, 7, 12, 13, 14, 19, 20, 15, 10, 5, 4, 9, 14, 15, 10, 9, 8, 3, 2, 7, 6, 11, 12, 17, 18, 17, 16, 11, 6, 1. There are 38 edges in this path and so the inspector must travel 3800 yards.

One approach to solving this problem is to add edges of minimal total length which make the graph Eulerian and then determining an Eulerian cycle. Inspection leads to the addition of redundant edges from 6 to 11, 17 to 18, 2 to 3, 10 to 15 and 4 to 19 guaranteeing every edge has even degree. The above path is an Eulerian cycle in this graph.

7. Recall $K_{2,2,2,2}$ can be constructed as four rows of two nodes with an edge connecting nodes in different rows. The graph below shows a subset of edges of $K_{2,2,2,2}$ which demonstrate $K_{2,2,2,2}$ contains a homeomorph of K_5. The nodes 1-5 correspond to those of K_5. Note the only "edges" of K_5 which require paths are those from 2 to 3 and 4 to 5.

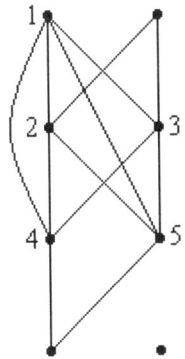

Section 4.3 Exercises

1.

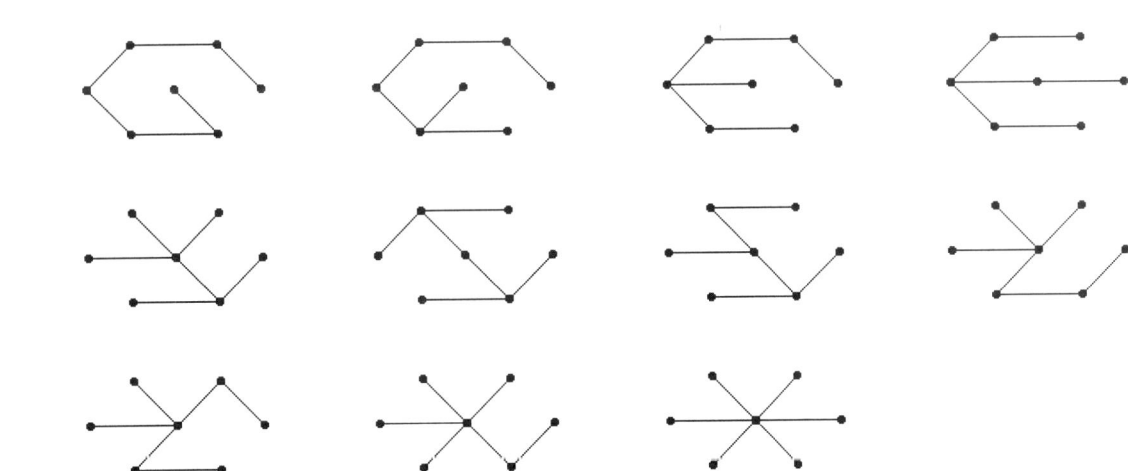

3.

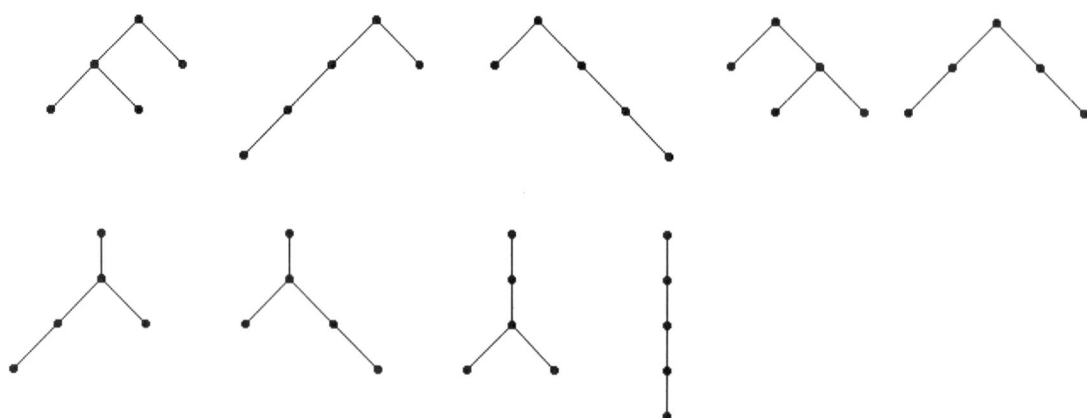

5. For the octahedron

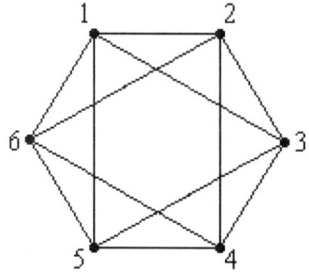

five spanning trees can be listed by their Prüfer codes: (2,1,6,5), (2,4,4,5), (1,1,5,1), (3,3,1,6), (2,4,5,1).

7. For the graph of the icosahedron (Figure 4.1.6 in the text), five spanning tress are pictured below.

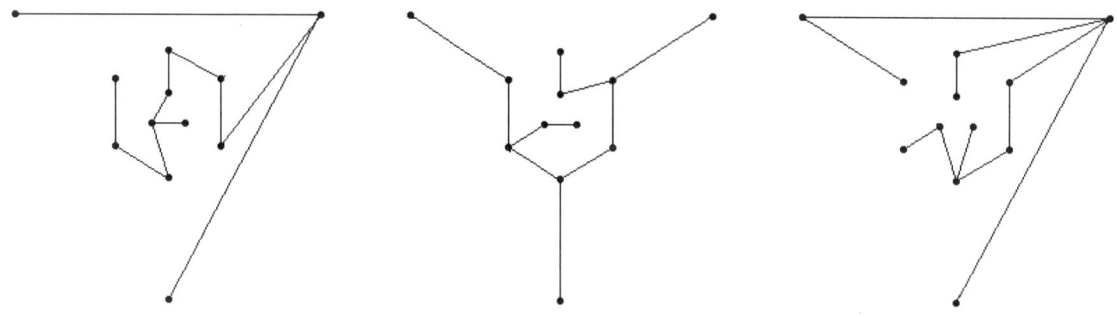

9. Number the nodes, in order from top to bottom, left to right, with the numbers 1, 2, 3, 4, 5, 6. The matrix of the Matrix-Tree Theorem is then

$$\begin{pmatrix} 2 & -1 & 0 & -1 & 0 & 0 \\ -1 & 3 & -1 & 0 & -1 & 0 \\ 0 & -1 & 2 & 0 & 0 & -1 \\ -1 & 0 & 0 & 2 & -1 & 0 \\ 0 & -1 & 0 & -1 & 3 & -1 \\ 0 & 0 & -1 & 0 & -1 & 2 \end{pmatrix}$$

which has cofactors equal to 15, the number of spanning trees.

11. Number the nodes as pictured.

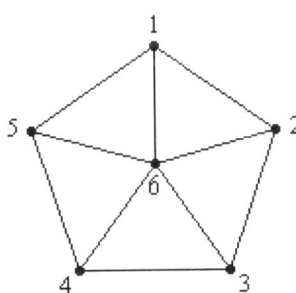

The matrix of the Matrix-Tree Theorem is then

$$\begin{pmatrix} 3 & -1 & 0 & 0 & -1 & -1 \\ -1 & 3 & -1 & 0 & 0 & -1 \\ 0 & -1 & 3 & -1 & 0 & -1 \\ 0 & 0 & -1 & 3 & -1 & -1 \\ -1 & 0 & 0 & -1 & 3 & -1 \\ -1 & -1 & -1 & -1 & -1 & 5 \end{pmatrix}$$

which has cofactors equal to 121, the number of spanning trees.

13. Number the nodes as pictured.

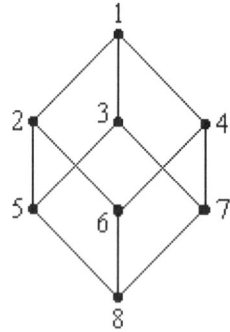

The matrix of the Matrix-Tree Theorem is then

$$\begin{pmatrix} 3 & -1 & -1 & -1 & 0 & 0 & 0 & 0 \\ -1 & 3 & 0 & 0 & -1 & -1 & 0 & 0 \\ -1 & 0 & 3 & 0 & -1 & 0 & -1 & 0 \\ -1 & 0 & 0 & 3 & 0 & -1 & -1 & 0 \\ 0 & -1 & -1 & 0 & 3 & 0 & 0 & -1 \\ 0 & -1 & 0 & -1 & 0 & 3 & 0 & -1 \\ 0 & 0 & -1 & -1 & 0 & 0 & 3 & -1 \\ 0 & 0 & 0 & 0 & -1 & -1 & -1 & 3 \end{pmatrix}$$

which has cofactors equal to 384, the number of spanning trees.

15.

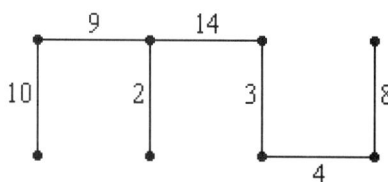

17. A breadth-first spanning tree is on the left, a depth-first spanning tree on the right.

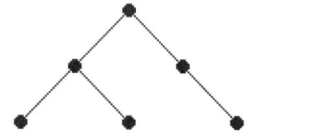 

19. Number the nodes of the given graph 1, 2, 3, ... starting at the top and moving left-to-right and top-to-bottom. The desired traversals are specified using this numbering.

inorder:	7, 4, 8, 2, 9, 5, 1, 10, 6, 11, 3
preorder:	1, 2, 4, 7, 8, 5, 9, 3, 6, 10, 11
postorder:	7, 8, 4, 9, 5, 2, 10, 11, 6, 3, 1

Section 4.3 Advanced Exercises

1. The twenty-three different unlabelled trees on eight nodes can be listed using Prüfer codes as follows assuming some numbering of nodes from 1 to 8.

(1,1,1,1,1,1)	(2,2,1,1,1,1)	(2,2,2,1,1,1)	(2,1,1,1,1,1)	(3,3,4,5,5,7)	(1,1,4,1,6,7)
(3,3,5,5,6,5)	(2,3,4,5,3,3)	(2,3,1,1,1,1)	(1,1,2,2,3,3)	(2,3,3,5,3,7)	(1,2,1,3,1,1)
(1,2,1,3,2,3)	(1,3,1,2,2,1)	(1,3,3,1,6,7)	(1,1,4,5,5,6)	(1,2,3,4,1,1)	(1,1,2,3,4,5)
(1,2,3,4,5,6)	(1,2,1,3,4,1)	(1,2,1,3,4,5)	(1,2,1,3,1,4)	(1,2,3,1,4,5)	

3. Assume the nodes of each rooted binary tree are numbered from 1 to 7 with the root labeled 7. Prüfer codes for the fifty-one rooted binary trees with 7 nodes are listed in the table below.

(5,5,6,6,7)	(2,3,5,5,7)	(4,4,5,6,6)	(3,3,5,5,6)	(2,5,5,6,7)	(2,4,5,5,7)
(4,5,5,6,6)	(4,4,5,5,6)	(3,5,5,6,7)	(2,3,6,6,7)	(2,4,4,6,6)	(3,4,5,5,6)
(4,5,5,6,7)	(2,4,6,6,7)	(3,4,4,6,6)	(2,3,5,5,6)	(2,5,6,6,7)	(2,3,5,6,7)
(2,5,5,6,6)	(2,4,5,5,6)	(3,5,6,6,7)	(2,4,5,6,7)	(3,5,5,6,6)	(2,4,4,5,6)
(4,5,6,6,7)	(2,4,4,5,7)	(2,4,5,6,6)	(3,4,4,5,6)	(3,4,5,5,7)	(3,4,4,5,7)
(3,4,5,6,6)	(3,3,4,5,6)	(3,3,5,5,7)	(3,4,4,6,7)	(3,3,4,6,6)	(2,3,4,5,6)
(4,4,5,5,7)	(2,4,4,6,7)	(3,3,5,6,6)	(3,4,6,6,7)	(3,3,4,5,7)	(2,3,4,6,6)
(4,4,6,6,7)	(3,3,4,6,7)	(2,3,5,6,6)	(3,3,6,6,7)	(2,3,4,5,7)	(3,3,5,6,7)
(2,3,4,6,7)	(4,4,5,6,7)	(3,4,5,6,7)			

5. Number the nodes of the 7-wheel by labeling the top node 1 and continue labeling nodes 2-7, in order, in a clockwise fashion. Label the center node 8. The matrix of the Matrix-Tree Theorem is then

$$\begin{pmatrix} 3 & -1 & 0 & 0 & 0 & 0 & -1 & -1 \\ -1 & 3 & -1 & 0 & 0 & 0 & 0 & -1 \\ 0 & -1 & 3 & -1 & 0 & 0 & 0 & -1 \\ 0 & 0 & -1 & 3 & -1 & 0 & 0 & -1 \\ 0 & 0 & 0 & -1 & 3 & -1 & 0 & -1 \\ 0 & 0 & 0 & 0 & -1 & 3 & -1 & -1 \\ -1 & 0 & 0 & 0 & 0 & -1 & 3 & -1 \\ -1 & -1 & -1 & -1 & -1 & -1 & -1 & 7 \end{pmatrix}$$

which has cofactors equal to 841, the number of spanning trees.

7. Label the nodes in the top row 1-5 and in the bottom row 6-10. The matrix of the Matrix-Tree Theorem is then

$$\begin{pmatrix} 2 & -1 & 0 & 0 & 0 & -1 & 0 & 0 & 0 & 0 \\ -1 & 3 & -1 & 0 & 0 & 0 & -1 & 0 & 0 & 0 \\ 0 & -1 & 3 & -1 & 0 & 0 & 0 & -1 & 0 & 0 \\ 0 & 0 & -1 & 3 & -1 & 0 & 0 & 0 & -1 & 0 \\ 0 & 0 & 0 & -1 & 2 & 0 & 0 & 0 & 0 & -1 \\ -1 & 0 & 0 & 0 & 0 & 2 & -1 & 0 & 0 & 0 \\ 0 & -1 & 0 & 0 & 0 & -1 & 3 & -1 & 0 & 0 \\ 0 & 0 & -1 & 0 & 0 & 0 & -1 & 3 & -1 & 0 \\ 0 & 0 & 0 & -1 & 0 & 0 & 0 & -1 & 3 & -1 \\ 0 & 0 & 0 & 0 & -1 & 0 & 0 & 0 & -1 & 2 \end{pmatrix}$$

which has cofactors equal to 209, the number of spanning trees.

9. The graph $K_{2,2,2,2}$ is constructed with four rows of 3 nodes, the top row numbered from 1-2, the second row 3-4, the third row 5-6 and the bottom row numbered 7-8. Each node in a row is connected to all nodes not in that row with an edge. Thus the matrix of the Matrix-Tree Theorem is

$$\begin{pmatrix} 6 & 0 & -1 & -1 & -1 & -1 & -1 & -1 \\ 0 & 6 & -1 & -1 & -1 & -1 & -1 & -1 \\ -1 & -1 & 6 & 0 & -1 & -1 & -1 & -1 \\ -1 & -1 & 0 & 6 & -1 & -1 & -1 & -1 \\ -1 & -1 & -1 & -1 & 6 & 0 & -1 & -1 \\ -1 & -1 & -1 & -1 & 0 & 6 & -1 & -1 \\ -1 & -1 & -1 & -1 & -1 & -1 & 6 & 0 \\ -1 & -1 & -1 & -1 & -1 & -1 & 0 & 6 \end{pmatrix}$$

which has cofactors equal to 82944, the number of spanning trees.

Section 4.4 Supplementary Exercises

1. 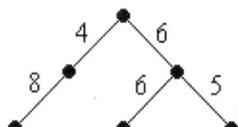   3.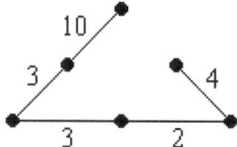

5. Breadth-first: 1, 2, 7, 3, 4, 5, 6   Depth-first: 1, 2, 3, 4, 7, 5, 6

In each of exercises 7 and 9, number the nodes of the given graph 1, 2, 3, ... starting at the top and moving left-to-right and top-to-bottom. The desired traversals are specified using this numbering.

7.   inorder: 4, 2, 5, 1, 6, 3, 7
     preorder: 1, 2, 4, 5, 3, 6, 7
     postorder: 4, 5, 2, 6, 7, 3, 1

9.   inorder: 4, 2, 7, 5, 1, 8, 6, 3
     preorder: 1, 2, 4, 5, 7, 3, 6, 8
     postorder: 4, 7, 5, 2, 8, 6, 3, 1

Chapter 5 Proof Techniques and Logic

Section 5.1 Exercises

1. Denote "The sentence on the other side of this paper is true." by P1 and "The sentence on the other side of this paper is false." by P2. Assume P1 is true. It follows that P2 is true since P1 declares it so. If P2 is true then P1 must be false since P2 declares it to be. Since P1 cannot be both true and false, there is a contradiction and so our original assumption is incorrect, P1 cannot be true. Now assume P2 is true. Then P1 is false, P2 says so. For P1 to be false, P2 must be false, again a contradiction. Therefore, neither P1 nor P2 can be true. An exactly similar argument shows neither P1 nor P2 can be false.

3. There are eighteen syllables in the phrase "the least integer not nameable in fewer than nineteen syllables." If this phrase described an integer then that integer would be nameable in fewer than nineteen syllables contradicting its definition. Hence no integer is described by the phrase.

5. $((p \wedge q) \to r)$ 7. $((p \vee r) \to q)$ 9. $((p \vee (\sim q)) \to (\sim r))$

11. Joan is not at the office.

13. If John or Laura is at the office then Joan is at the office.

15. If, whenever Joan is not at the office Laura is at the office, then John is at the office.

17. John is not at the office and either Joan or Laura is at the office.

19. Let m be an integer and assume m is not divisible by 5. Then m has one of the forms $5k+1$, $5k+2$, $5k+3$ or $5k+4$ for some integer k. The four cases and the corresponding value of m^2 are
listed below:

$m = 5k+1$ $m^2 = 25k^2 + 10k + 1 = 5(5k^2 + 2k) + 1$

$m = 5k+2$ $m^2 = 25k^2 + 20k + 4 = 5(5k^2 + 4k) + 4$

$m = 5k+3$ $m^2 = 25k^2 + 30k + 9 = 5(5k^2 + 6k + 1) + 4$

$m = 5k+4$ $m^2 = 25k^2 + 40k + 16 = 5(5k^2 + 8k + 3) + 1$

In all cases m^2 is a multiple of 5 plus either 1 or 4 and so m^2 is not divisible by 5.

21. We follow the same steps as in exercise 20. Assume $\sqrt{10} = \frac{m}{n}$ with m and n having no common factors. Then $m^2 = 10n^2$ and so $m = 2k$ is even. It follows that $10n^2 = (2k)^2 = 4k^2$ or $5n^2 = 2k^2$. Hence n^2 and n are even contradicting the fact m and n have no common factors. Therefore $\sqrt{10}$ is irrational.

23. Assume $\sqrt{5}$ is rational and write $\sqrt{5} = \frac{m}{n}$ where $\frac{m}{n}$ is in lowest terms. Then $m^2 = 5n^2$ and m^2 is divisible by 5. It follows from exercise 19 that $m = 5k$ for some integer k. Thus $5n^2 = (5k)^2 = 25k^2$ or $n^2 = 5k^2$. Therefore n^2 is divisible by 5 and, again applying exercise 19, n is divisible by 5. Since m and n both have a common factor of 5, $\frac{m}{n}$ is not in lowest terms as we assumed. Therefore $\sqrt{5}$ is irrational by contradiction.

25. The proof that (ii) implies (i) is contained in exercise 18 so to prove logical equivalence we must show (i) implies (ii). Assume m is an integer divisible by 3. Then $m = 3k$ for some integer k. It follows that $m^2 = (3k)^2 = 9k^2 = 3(3k^2)$ and so m^2 must be divisible by 3.

27.   (i) implies (ii). See exercise 25.
    (ii) implies (iii). Assume m^2 is divisible by 3. Then $m^2 = 3k$ for some integer k and $m^3 = m(m^2) = m(3k) = 3(mk)$ and so m^3 is divisible by 3.
    (iii) implies (i). Assume m^3 is divisible by 3. The integer m must have one of the forms 3k, $3k+1$, or $3k+2$. If $m = 3k+1$ then $m^3 = 27k^3 + 27k^2 + 9k + 1 = 3(9k^3 + 9k^2 + 3k) + 1$ which is one more than a multiple of 3 contradicting the assumption that m^3 is divisible by 3. Similarly, $m = 3k+2$ implies $m^3 = 27k^3 + 54k^2 + 36k + 8 = 3(9k^3 + 18k^2 + 12k + 2) + 2$ again contradicting the assumption. It follows that $m = 3k$ and so m is divisible by 3.

Section 5.1 Advanced Exercises

1. Two questions are needed as the traveler now needs to eliminate two roads as possibilities. First the traveler points to the road on the right asks, "If you belonged to the other tribe would you say that this was the road to the city?" As argued in the text, a response of "no" means the traveler should take the right fork. An answer of "yes" eliminates the road to the right as a possibility but now the traveler must determine which of the middle and left roads to take. The traveler points to the leftmost road and repeats his question. If the response is "no", the traveler takes the left road, if "yes" he takes the middle road.

Section 5.2 Exercises

1. Universal affirmative. Code letter: A. The set of Cretans is a subset of the set of liars.

3. Particular affirmative. Code letter: I. Particular affirmative. Code letter: I. The intersection of the set of members of the Committee on Bylaws and the set of members of the Committee on Committees is non-empty.

5. Universal negative. Code letter: E. The intersection of the set of members of the Rank and Tenure Committee and the set of members of the Committee on Bylaws is empty.

7. Particular negative. Code letter: O. The intersection of the set of members of the Committee on Committees and the set of people who are not members of the Committee on Bylaws is non-empty.

9. Statements (i)-(iii) are all of type Universal Affirmative and so the syllogism has code AAA. Let C denote the set of Cretans and L the set of liars. Statements (i) and (ii) translate to $C \subset L$ and $\{Epimenides\} \subset C$ respectively. Hence $\{Epimenides\} \subset L$ which is the set equivalent of statement (iii).

11. Statements (i)-(iii) are all of type Universal Affirmative and so the syllogism has code AAA. Let D denote the set of Danish villains and K the set of arrant knaves. Statements (i) and (ii) translate to $D \subset K$ and $\{Claudius\} \subset D$ respectively. Hence $\{Claudius\} \subset K$ which is the set equivalent of statement (iii).

13. Suppose the universe is the set of all stories. Let S denote a set containing one element, your story about how you once met a sea serpent. Let Y be the set of stories which make me yawn and D the stories totally devoid of interest. Statement (i) is of type Universal Affirmative $(S \subset Y)$ as is statement (ii) $(Y \subset D)$. It follows immediately that $S \subset D$ which is the Universal Affirmative of statement (iii). The syllogism code is AAA.

15. The argument can be factored into three syllogisms:

    (i) Some members of the faculty are wealthy.
    (ii) All wealthy faculty are members of the Committee on Committees.
    (iii) Some members of the Committee on Committees are wealthy faculty.

    (iii) Some members of the Committee on Committees are wealthy faculty.
    (iv) All wealthy faculty who are members of the Committee on Committees are also members of the
        Secret Subcommittee of the Committee on Committees.
    (v) There are wealthy faculty on the Secret Subcommittee of the Committee on Committees.

    (v) There are wealthy faculty on the Secret Subcommittee of the Committee on Committees.
    (vi) No one can be elected to the Secret Subcommittee of the Committee on Committees unless he or
        she is a senior faculty member. (All members of the Secret Subcommittee of the Committee are
        senior faculty members.)
    (vii) Some senior faculty are wealthy.

The AEIO code for all three of these syllogisms is IAI and the code for the entire argument is IAIIAIIAI.

Section 5.2 Advanced Exercises

1. Begin by rewriting each of the 12 statements in more traditional forms.

    (i)   All "easterly wind" times are "sunny" times.
    (ii)   All "cold and foggy" times are "flute practicing" times.
    (iii)   All "fire smoking" times are "open door" times.
    (iv)   All "cold and rheumatic" times are "fire" times.
    (v)   All "easterly wind and gusty" times are "fire smoking" times.

(vi)   All "open door" times are "headache free" times.
(vii) All "sunny and foggy and not cold" times are "shut window" times.
(viii) All "rheumatic times" are "gusty" or "no fire" or "open door" times.
(ix)   All "sunny" times are "foggy" times.
(x)   All "open door" times are "headache" or "no flute practicing" times.
(xi)   All "easterly wind and foggy" times are "rheumatic" times.
(xii) All "easterly wind" times are "shut window" times.

Note that because of statement (vi), statement (x) can be simplified to

(x)   No "flute practicing" times are "open door" times.

The argument can now be factored into ten syllogisms as follows.

(i)   All "easterly wind" times are "sunny" times.
(ix) All "sunny" times are "foggy" times.
C1:   All "easterly wind" times are "foggy" times.

(xi*) All "easterly wind" times are "rheumatic" times.   (C1 applied to (xi))
(viii) All "rheumatic times" are "gusty" or "no fire" or "open door" times.
(ix)   C2:   All "easterly wind" times are "gusty" or "no fire" or "open door" times.

(v)   All "easterly wind and gusty" times are "fire smoking" times.
(iii) All "fire smoking" times are "open door" times.
C3:   All "easterly wind and gusty" times are "open door" times.

C3   All "easterly wind and gusty" times are "open door" times.
(x)   No "flute practicing" times are "open door" times.
C4:   No "flute practicing" times are "easterly wind and gusty" times.

C4   No "flute practicing" times are "easterly wind and gusty" times.
(ii) All "cold and foggy" times are "flute practicing" times.
C5:   No "cold and foggy" times are "easterly wind and gusty" times.

Using C1 and (i) and C5 above, we can rewrite C5 as

C5:   All "easterly wind and gusty" times are "sunny and foggy and not cold" times.
(vii) All "sunny and foggy and not cold" times are "shut window" times.
C6:   All "easterly wind and gusty" times are "shut window" times.

Combining (iv) and (xi*) we know:

   All "easterly wind and no fire" times are "not cold" times.

Combining C1 and (i) we get (xiii)

(xii) All "easterly wind and no fire" times are "sunny and foggy and not cold" times.
(vi)   All "sunny and foggy and not cold" times are "shut window" times.
C7:   All "easterly wind and no fire" times are "shut window" times.

       All "easterly wind and open door" times are "open door" times.   (obvious)
(x)   No "flute practicing" times are "open door" times.
C8:   No "flute practicing" times are "easterly wind and open door" times.

60 Chapter 5 Proof Techniques and Logic

C7   No "flute practicing" times are "easterly wind and open door" times.
(ii)   All "cold and foggy" times are "flute practicing" times.
C9:   No "cold and foggy" times are "easterly wind and open door" times.

Using C1 and (i) and C8 above, we can rewrite C8 as

C9:   All "easterly wind and open door" times are "sunny and foggy and not cold" times.
(viii) All "sunny and foggy and not cold" times are "shut window" times.
C10:   All "easterly wind and open door" times are "shut window" times.

Combining C2, C6, C7 and C10 gives the desired conclusion. The code of the above syllogisms is:

AAAAAAAAAEEEAEAAAAAAEEEAEAAA

Section 5.3 Exercises

1. Well-formed

3. Not well-formed, one too many right parentheses.

5. Not well-formed, need a closing right parenthesis at the end.

7. $((p \vee q) \rightarrow (p \wedge q))$ is not a tautology.

p	q	$(p \vee q)$	$(p \wedge q)$	$((p \vee q) \rightarrow (p \wedge q))$
T	T	T	T	T
T	F	T	F	F
F	T	T	F	F
F	F	F	F	T

9. $((p \rightarrow q) \leftrightarrow (p \vee q))$ is not a tautology.

p	q	$(p \rightarrow q)$	$(p \vee q)$	$((p \rightarrow q) \leftrightarrow (p \vee q))$
T	T	T	T	T
T	F	F	T	F
F	T	T	T	T
F	F	T	F	F

11. $((\sim (p \vee q)) \leftrightarrow ((\sim p) \wedge (\sim q)))$ is a tautology.

p	q	(~ (p ∨ q))	(~ p)	(~ q)	((~ p) ∧ (~ q))	((~ (p ∨ q)) ↔ ((~ p) ∧ (~ q)))
T	T	F	F	F	F	T
T	F	F	F	T	F	T
F	T	F	T	F	F	T
F	F	T	T	T	T	T

13.

p	(~ (~ p))	((~ (~ p)) ↔ p)
T	T	T
F	F	T

15.

p	q	r	(q ∨ r)	(p ∨ (q ∨ r))	(p ∨ q)	((p ∨ q) ∨ r)	((p ∨ (q ∨ r)) ↔ ((p ∨ q) ∨ r)
T	T	T	T	T	T	T	T
T	F	T	T	T	T	T	T
T	T	F	T	T	T	T	T
T	F	F	F	T	T	T	T
F	T	T	T	T	T	T	T
F	F	T	T	T	F	T	T
F	T	F	T	T	T	T	T
F	F	F	F	F	F	F	T

p	q	r	(q ∧ r)	(p ∧ (q ∧ r))	(p ∧ q)	((p ∧ q) ∧ r)	((p ∧ (q ∧ r)) ↔ ((p ∧ q) ∧ r)
T	T	T	T	T	T	T	T
T	F	T	F	F	F	F	T
T	T	F	F	F	T	F	T
T	F	F	F	F	F	F	T
F	T	T	T	F	F	F	T
F	F	T	F	F	F	F	T
F	T	F	F	F	F	F	T
F	F	F	F	F	F	F	T

17.

p	q	(p ∨ q)	(~ p)	((~ p) → q)	((p ∨ q) ↔ ((~ p) → q)
T	T	T	F	T	T
T	F	T	F	T	T
F	T	T	T	T	T
F	F	F	T	F	T

19. The first of DeMorgan's law was verified in exercise 11. The second is verified by the truth table below.

p	q	$(\sim(p \wedge q))$	$(\sim p)$	$(\sim q)$	$((\sim p) \vee (\sim q))$	$((\sim(p \wedge q)) \leftrightarrow ((\sim p) \vee (\sim q)))$
T	T	F	F	F	F	T
T	F	T	F	T	T	T
F	T	T	T	F	T	T
F	F	T	T	T	T	T

21. No, $(((p \wedge q) \to r) \leftrightarrow ((p \to q) \wedge (q \to r)))$ is not a logical identity. If p, q, and r have truth values T, F and T respectively then $((p \to q) \wedge (q \to r))$ has truth value F while $((p \wedge q) \to r)$ has truth value T.

23. Yes, $((p \to q) \vee (q \to p))$ is a tautology. For this formula to be false, both terms of the disjunction must be false. For $(p \to q)$ to be false, p must be true and q false, however then $(q \to p)$ is true. Thus one term must always be true.

Section 5.3 Advanced Exercises

7. There are 42 ways to parenthesize six letters as given in the table below.

$a(b(c(d(ef))))$	$a(b(c((de)f)))$	$a(b((cd)(ef)))$	$a(b(((cd)e)f))$	$a(b((c(de))f))$	$a((bc)((de)f))$
$a((bc)(d(ef)))$	$a(((bc)d)(ef))$	$a((b(cd))(ef))$	$a((b(c(de)))f)$	$a((b((cd)e))f)$	$a(((bc)(de))f)$
$a((((bc)d)e)f)$	$a(((b(cd))e)f)$	$(ab)(c(d(ef)))$	$(ab)(c((de)f))$	$(ab)((cd)(ef))$	$(ab)(((cd)e)f)$
$(ab)((c(de))f)$	$((ab)c)((de)f)$	$((ab)c)(d(ef))$	$(a(bc))((de)f)$	$(a(bc))(d(ef))$	$(a(b(cd)))(ef)$
$(a((bc)d))(ef)$	$((ab)(cd))(ef)$	$(((ab)c)d)(ef)$	$((a(bc))d)(ef)$	$(a(b(c(de))))f$	$(a(b((cd)e)))f$
$(a((bc)(de)))f$	$(a(((bc)d)e))f$	$(a((b(cd))e))f$	$((ab)((cd)e))f$	$((ab)(c(de)))f$	$(((ab)c)(de))f$
$((a(bc))(de))f$	$((a(b(cd)))e)f$	$((a((bc)d))e)f$	$(((ab)(cd))e)f$	$((((ab)c)d)e)f$	$(((a(bc))d)e)f$

Section 5.4 Supplementary Exercises

1. The syllogism is (i) All B is A. (ii) Some C is B. (iii) Some C is A.
In terms of sets, (i) says $B \subset A$ and (ii) says C intersects B. Therefore C must intersect A which is the set interpretation of (iii).

3. The syllogism is (i) No B is A. (ii) Some C is B. (iii) Some C is not A.
In terms of sets, (i) states A and B do not intersect while (ii) says C intersects B. The elements in $B \cap C$ cannot be in A by (i) so $C \cap A^c \neq \emptyset$ which is the set interpretation of (iii).

5. The syllogism is (i) No A is B. (ii) All C is B. (iii) No C is A.
In terms of sets, (i) states $A \cap B = \emptyset$ while (ii) says C is contained in B. Any element of C lives in B and so cannot be in A by (i) so $C \cap A = \emptyset$ which is the set interpretation of (iii).

7. The syllogism is (i) No B is A. (ii) Some B is C. (iii) Some C is not A.
In terms of sets, (i) states $A \cap B = \emptyset$ while (ii) says $C \cap B \neq \emptyset$. Any element of $C \cap B$ is in B and cannot be in A by (i) so $C \cap A^c \neq \emptyset$ which is the set interpretation of (iii).

9. The syllogism is (i) Some B is not A. (ii) All B is C. (iii) Some C is not A.
In terms of sets, (i) states $B \cap A^c \neq \emptyset$ while (ii) says $B \subset C$. Any element of $B \cap A^c$ is in B and so in C by (ii) so $C \cap A^c \neq \emptyset$ which is the set interpretation of (iii).

11. The syllogism is (i) No A is B. (ii) Some B is C. (iii) Some C is not A.
In terms of sets, (i) states A and B do not intersect while (ii) says C intersects B. The elements in $B \cap C$ cannot be in A by (i) so $C \cap A^c \neq \emptyset$ which is the set interpretation of (iii).

13. The syllogism is (i) All B is A. (ii) Some C is not B. (iii) Some C is not A. Let $A = C = \{1,2\}$ and $B = \{1\}$ then (i) and (ii) are true statements but (iii) is not a true statement.

15. The syllogism is (i) All A is B. (ii) Some C is B. (iii) Some C is A. Let $A = \{1,2\}$, $B = \{1,2,3,4\}$ and $C = \{3,4\}$ then (i) and (ii) are true statements but (iii) is not a true statement.

17. The syllogism is (i) All B is A. (ii) All B is C. (iii) All C is A. Let $A = \{1,2,3\}$, $B = \{1,2\}$ and $C = \{1,2,4\}$ then (i) and (ii) are true statements but (iii) is not a true statement.

19. The syllogism is (i) All A is B. (ii) Some B is not C. (iii) Some C is not A. Let $A = \{1,2\}$, $B = \{1,2,3\}$ and $C = \{1\}$ then (i) and (ii) are true statements but (iii) is not a true statement.

Chapter 6 Boolean Algebras, Boolean Functions, and Logic

Section 6.1 Exercises

1.
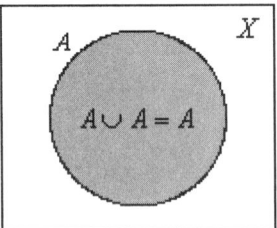

3.
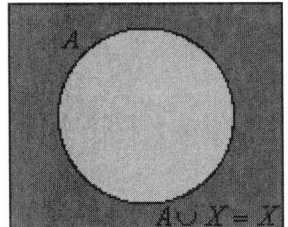

5.
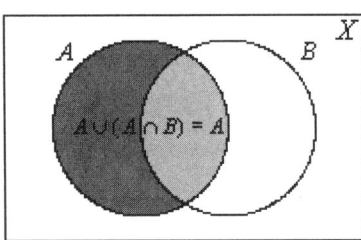

7. Not a Boolean algebra, $\{a\} \cup \{b,c\} \notin X$ for example.

9. Not a Boolean algebra, $\{a\}^C = \{b,c\} \notin X$ for example.

11. Not a Boolean algebra, $\{a\}^C = \{b,c\} \notin X$ for example.

13. Define $f(x,y) = x + y$ then
    $f(0,0) = 0 \quad f(1,0) = 1 \quad f(0,1) = 1 \quad f(1,1) = 1$.

15. Define $f(x,y) = xy$ then
    $f(0,0) = 0 \quad f(1,0) = 0 \quad f(0,1) = 0 \quad f(1,1) = 1$.

Section 6.1 Boolean Algebras and Functions   65

17. Define $f(x,y,z) = x + y + z$ then
    $f(0,0,0) = 0$ $f(1,0,0) = 1$ $f(0,1,0) = 1$ $f(0,0,1) = 1$
    $f(1,1,0) = 1$ $f(1,0,1) = 1$ $f(0,1,1) = 1$ $f(1,1,1) = 1$.

19. Define $f(x,y,z) = \overline{x} + yz$ then
    $f(0,0,0) = 1$ $f(1,0,0) = 0$ $f(0,1,0) = 1$ $f(0,0,1) = 1$
    $f(1,1,0) = 0$ $f(1,0,1) = 0$ $f(0,1,1) = 1$ $f(1,1,1) = 1$.

21. Disjunctive normal form: $xy + x\overline{y} + \overline{x}y$. Conjunctive normal form: $x + y$.

23. Disjunctive normal form: $xyz + \overline{x}yz + x\overline{y}z + x\overline{y}\overline{z} + \overline{x}\overline{y}z$.
    Conjunctive normal form: $(x + y + z)(\overline{x} + y + z)(\overline{x} + \overline{y} + z)$.

25. Disjunctive normal form: $xyz + xy\overline{z} + x\overline{y}z + \overline{x}yz$.
    Conjunctive normal form: $(x + y + z)(\overline{x} + y + z)(x + \overline{y} + z)(x + y + \overline{z})$.

27. $(p \vee q)$

29. $(p \vee ((q \wedge r) \vee (p \wedge (\neg r))))$

Section 6.1 Advanced Exercises

1. $x + (y + z) = 1(x + (y + z)) = (x + \overline{x})(x + (y + z)) = [x(x + (y + z))] + [\overline{x}(x + (y + z))]$
The first term in brackets is equal to x by Theorem 6.1.3 while the second term simplifies to
$\overline{x}(x + (y + z)) = \overline{x}x + \overline{x}(y + z) = 0 + \overline{x}(y + z) = \overline{x}(y + z)$. Thus $x + (y + z) = x + \overline{x}(y + z)$.
Similarly
$(x + y) + z = 1((x + y) + z) = (x + \overline{x})((x + y) + z) = [x((x + y) + z)] + [\overline{x}((x + y) + z)]$
$= [x(x + y) + xz] + [\overline{x}(x + y) + \overline{x}z] = (x + xy) + [(\overline{x}x + \overline{x}y) + \overline{x}z]$
$= x + [(0 + \overline{x}y) + \overline{x}z] = x + (\overline{x}y + \overline{x}z) = x + \overline{x}(y + z)$
Thus $x + (y + z) = x + \overline{x}(y + z) = (x + y) + z$.

3. By Axiom 4 (and Axiom 1) we have $\overline{x} + x = x + \overline{x} = 1$ and $\overline{x} \cdot x = x \cdot \overline{x} = 0$. Thus x possesses the defining properties of $\overline{\overline{x}}$. By Exercise 2 above $\overline{\overline{x}} = x$.

5. To show $\overline{xy} = \overline{x} + \overline{y}$ we must show $\overline{x} + \overline{y}$ satisfies the defining properties stated in Axiom 4 with respect to the element xy. That is, we must show $xy + (\overline{x} + \overline{y}) = 1$ and $xy(\overline{x} + \overline{y}) = 0$. Throughout we freely use the associativity of addition (given by Exercise by 1) and multiplication (given by Exercise 1 and the Principle of Duality.) We have

$xy + (\overline{x} + \overline{y}) = (xy + \overline{x}) + \overline{y} = [xy + (\overline{x}y + \overline{x})] + \overline{y}$ (by Theorem 6.1.3)
$= [(x + \overline{x})y + \overline{x}] + \overline{y} = (1 \cdot y + \overline{x}) + \overline{y} = \overline{x} + (y + \overline{y}) = \overline{x} + 1 = 1$ (by Theorem 6.1.2.)

Also

$$xy(\overline{x}+\overline{y}) = (xy)\overline{x} + (xy)\overline{y} = (yx)\overline{x} + (xy)\overline{y} = y(x\overline{x}) + x(y\overline{y}) = y0 + x0 = 0 + 0 = 0$$

again using Theorem 6.1.2. Thus $\overline{xy} = \overline{x} + \overline{y}$.

Section 6.2 Exercises

1.

x	y	f(x,y)
0	0	0
0	1	1
1	0	1
1	1	1

(x,y)	f(x,y)
(0, 0)	0
(0, 1)	1
(1, 1)	1
(1, 0)	1

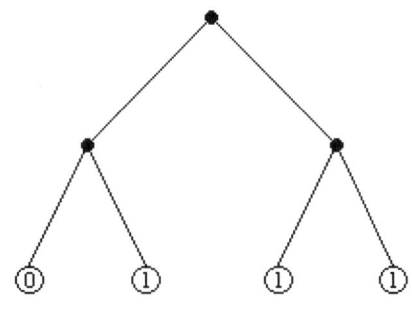
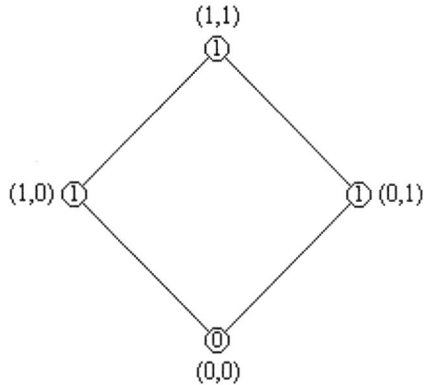

3.

x	y	f(x,y)
0	0	0
0	1	0
1	0	1
1	1	0

(x,y)	f(x,y)
(0, 0)	0
(0, 1)	0
(1, 1)	0
(1, 0)	1

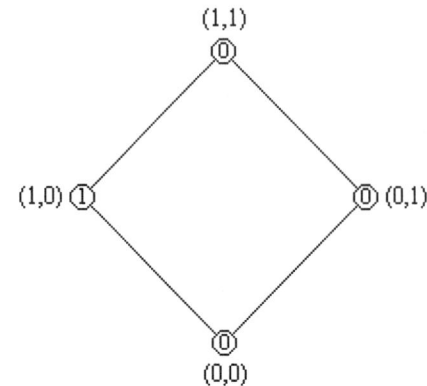
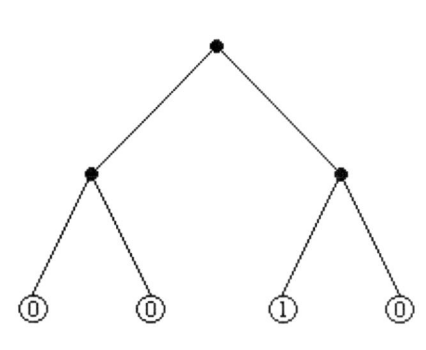

5.

x	y	z	f(x,y,z)
0	0	0	1
0	0	1	0
0	1	0	1
0	1	1	1
1	0	0	1
1	0	1	1
1	1	0	1
1	1	1	1

(x,y,z)	f(x,y,z)
(0, 0, 0)	1
(0, 0, 1)	0
(0, 1, 1)	1
(0, 1, 0)	1
(1, 1, 0)	1
(1, 1, 1)	1
(1, 0, 1)	1
(1, 0, 0)	1

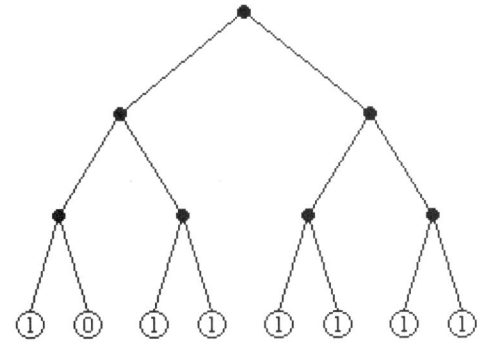
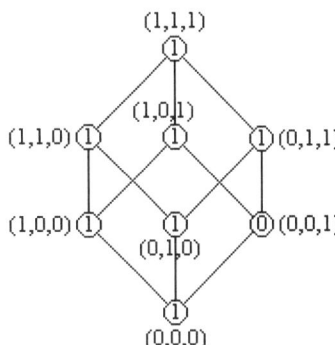

7.

x	y	z	f(x,y,z)
0	0	0	1
0	0	1	1
0	1	0	1
0	1	1	1
1	0	0	0
1	0	1	0
1	1	0	0
1	1	1	1

(x,y,z)	f(x,y,z)
(0, 0, 0)	1
(0, 0, 1)	1
(0, 1, 1)	1
(0, 1, 0)	1
(1, 1, 0)	0
(1, 1, 1)	1
(1, 0, 1)	0
(1, 0, 0)	0

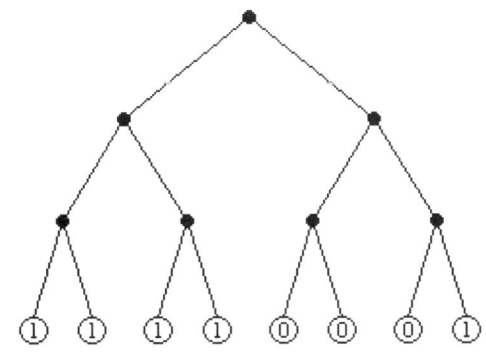
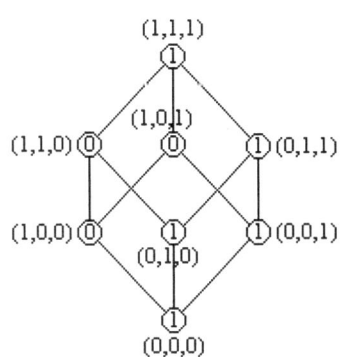

9.

x	y	z	w	f(x,y,z,w)
0	0	0	0	0
0	0	0	1	0
0	0	1	0	0
0	0	1	1	1
0	1	0	0	0
0	1	0	1	1
0	1	1	0	1
0	1	1	1	1
1	0	0	0	0
1	0	0	1	1
1	0	1	0	1
1	0	1	1	1
1	1	0	0	1
1	1	0	1	1
1	1	1	0	1
1	1	1	1	1

(x,y,z,w)	f(x,y,z,w)
(0, 0, 0, 0)	0
(0, 0, 0, 1)	0
(0, 0, 1, 1)	1
(0, 0, 1, 0)	0
(0, 1, 1, 0)	1
(0, 1, 1, 1)	1
(0, 1, 0, 1)	1
(0, 1, 0, 0)	0
(1, 1, 0, 0)	1
(1, 1, 0, 1)	1
(1, 1, 1, 1)	1
(1, 1, 1, 0)	1
(1, 0, 1, 0)	1
(1, 0, 1, 1)	1
(1, 0, 0, 1)	1
(1, 0, 0, 0)	0

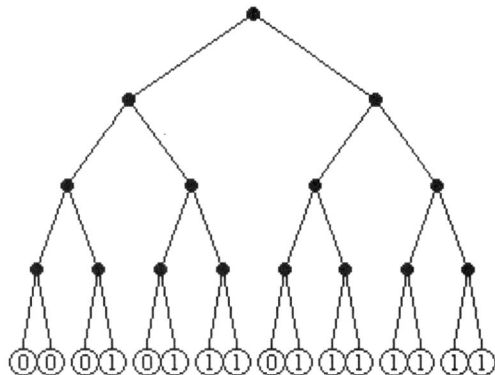

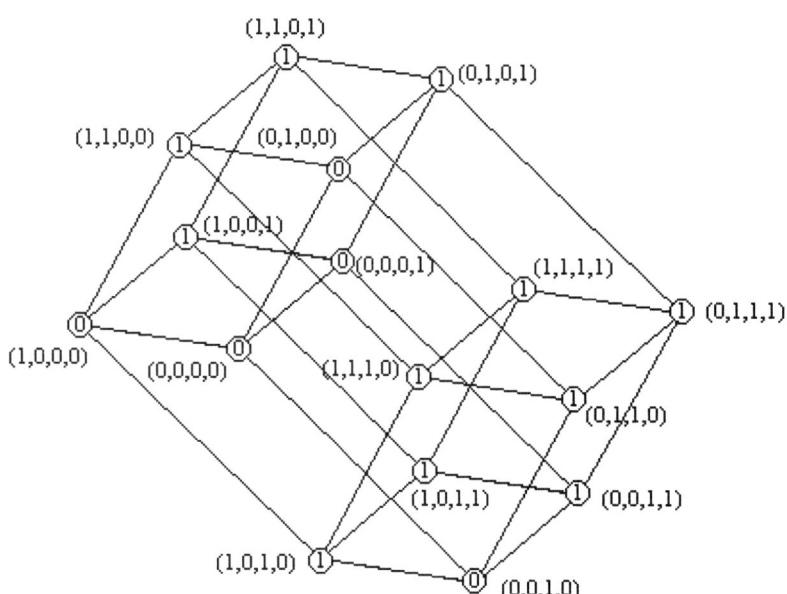

11. $f(x,y) = xy + x\bar{y} + \bar{x}y$

13. $f(x,y) = x\bar{y}$

15. $f(x,y,z) = xyz + \bar{x}yz + x\bar{y}z + xy\bar{z} + \bar{x}\bar{y}z + \bar{x}y\bar{z} + \bar{x}\bar{y}\bar{z}$

17. $f(x,y,z) = \bar{x}yz + \bar{x}\bar{y}z + \bar{x}y\bar{z} + x\bar{y}\bar{z} + xyz$

19. $f(x,y,z,w) = xyzw + xyz\bar{w} + xy\bar{z}w + \bar{x}y\bar{z}w + x\bar{y}zw + \bar{x}\bar{y}zw + \bar{x}y\bar{z}w + \bar{x}yzw +$
$x\bar{y}\bar{z}w + \bar{x}\bar{y}\bar{z}w + \bar{x}yz\bar{w}$

21. Using the Karnaugh map below, we have $xy + x\bar{y} = x(y+\bar{y}) = x$, $xy + \bar{x}y = (x+\bar{x})y = y$ so $f(x,y) = x + y$.

	y	$\bar{y}$
x	1	1
$\bar{x}$	1	0

23. $f(x,y) = x\bar{y}$ is already simplified. The corresponding Karnaugh map is below

	y	$\bar{y}$
x	0	1
$\bar{x}$	0	0

25. Using the Karnaugh map below, we have $xyz + xy\bar{z} + x\bar{y}\bar{z} + x\bar{y}z = x(y+\bar{y})(z+\bar{z}) = x$, $xyz + \bar{x}yz + xy\bar{z} + \bar{x}y\bar{z} = (x+\bar{x})y(z+\bar{z}) = y$ and $xy\bar{z} + x\bar{y}\bar{z} + \bar{x}y\bar{z} + \bar{x}\bar{y}\bar{z} = (x+\bar{x})(y+\bar{y})\bar{z} = \bar{z}$ so $f(x,y,z) = x + y + \bar{z}$.

	yz	$y\bar{z}$	$\bar{y}\bar{z}$	$\bar{y}z$
x	1	1	1	1
$\bar{x}$	1	1	1	0

27. Using the Karnaugh map below, we have $xyz + \bar{x}yz = (x+\bar{x})yz = yz$ and $\bar{x}yz + \bar{x}y\bar{z} + \bar{x}\bar{y}\bar{z} + \bar{x}\bar{y}z = \bar{x}(y+\bar{y})(z+\bar{z}) = \bar{x}$ so $f(x,y,z) = \bar{x} + yz$.

	yz	y$\bar{z}$	$\bar{y}\bar{z}$	$\bar{y}z$
x	1	0	0	0
$\bar{x}$	1	1	1	1

29. Using the Karnaugh map below, we have $xwyz + xwy\bar{z} + xw\bar{y}\bar{z} + xw\bar{y}z = xw(y+\bar{y})(z+\bar{z}) = xw$

$xwyz + \bar{x}wyz + x\bar{w}yz + \bar{x}\bar{w}yz = (x+\bar{x})(w+\bar{w})yz = yz$

$xwyz + xwy\bar{z} + \bar{x}wy\bar{z} + \bar{x}wy\bar{z} = (x+\bar{x})wy(z+\bar{z}) = wy$

$xwyz + xw\bar{y}z + \bar{x}wyz + \bar{x}w\bar{y}z = (x+\bar{x})w(y+\bar{y})z = wz$

$xwyz + xw\bar{y}\bar{z} + x\bar{w}yz + x\bar{w}y\bar{z} = x(w+\bar{w})y(z+\bar{z}) = xy$

$x\bar{w}yz + x\bar{w}\bar{y}z = x\bar{w}z$

so $f(x,y,z,w) = xw + yz + wy + wz + xy + x\bar{w}z$.

	yz	y$\bar{z}$	$\bar{y}\bar{z}$	$\bar{y}z$
xw	1	1	1	1
x$\bar{w}$	1	1	0	1
$\bar{x}\bar{w}$	1	0	0	0
$\bar{x}w$	1	1	0	1

In Exercises 31-39, the logic circuit corresponding to the sum of products form of *f* is on the left, the logic circuit for the simplified form of *f* is on the right. In cases where the two circuits are identical, only one circuit is given.

31.

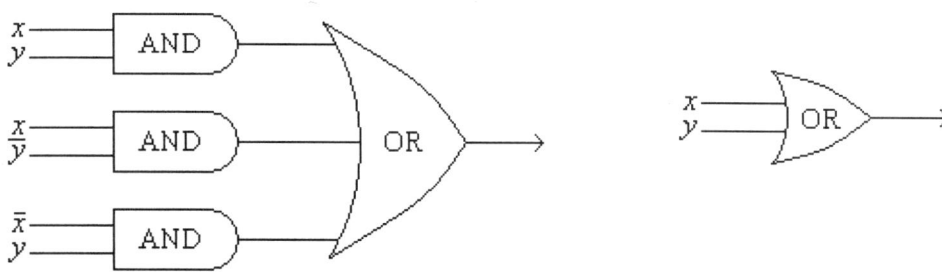

33.

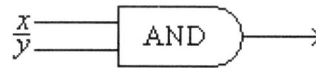

35.

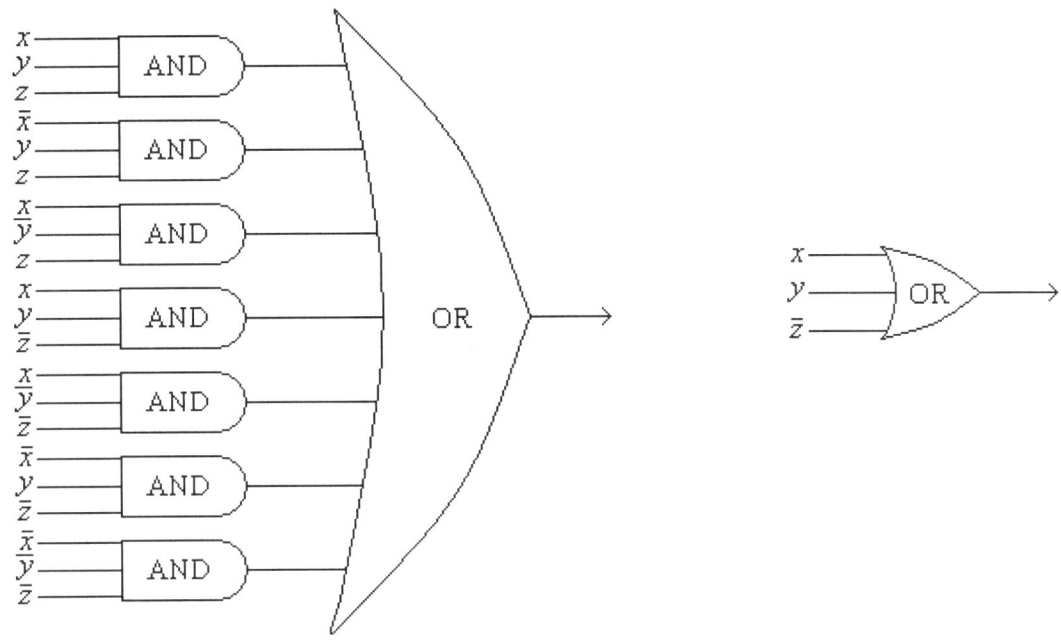

37.

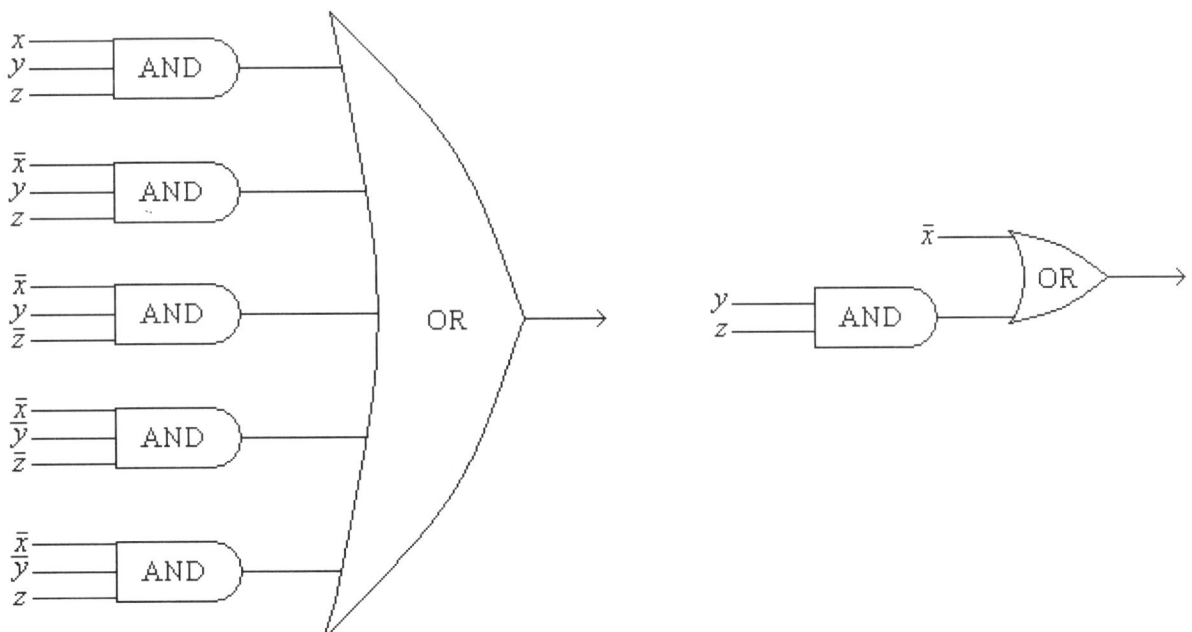

39. The logic circuit for the sum of products form of f is an OR gate having eleven AND gates as inputs. Each AND gate has four inputs, the first being x or $\bar{x}$, the second y or $\bar{y}$, the third z or $\bar{z}$, and the fourth w or $\bar{w}$. The eleven AND gates correspond to the terms

$\bar{x}yz\bar{w}$, $xy\bar{z}\bar{w}$, $\bar{x}y\bar{z}\bar{w}$, $\bar{x}\bar{y}z\bar{w}$, $x\bar{y}z\bar{w}$, $x\bar{y}z\bar{w}$, $\bar{x}\bar{y}z\bar{w}$, $\bar{x}yz\bar{w}$, $\bar{x}\bar{y}\bar{z}\bar{w}$ and $\bar{x}\bar{y}\bar{z}\bar{w}$. The simplified logic circuit for f appears below.

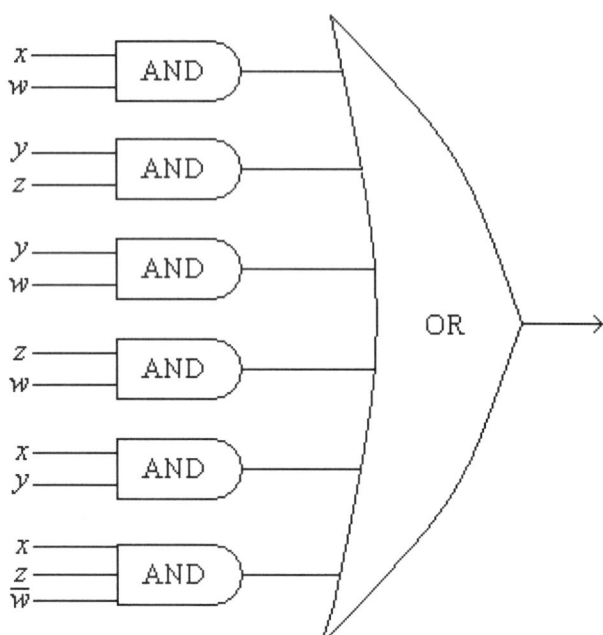

Section 6.2 Advanced Exercises

1. NAND(NAND(NAND(NAND(x,y),NAND(x,y)), NAND(NAND(x,y),NAND(x,y))), NAND(NAND(NAND(NAND(x,x),y),NAND(NAND(x,x),y)), NAND(NAND(NAND(x,x),y),NAND(NAND(x,x),y))))

NOR(NOR(NOR(NOR(x,x),NOR(y,y)),
NOR(NOR(NOR(x,x),NOR(x,x)),NOR(y,y))),NOR(NOR(NOR(x,x),NOR(y,y)),
NOR(NOR(NOR(x,x),NOR(x,x)),NOR(y,y))))

3. NAND(NAND(NAND(NAND(NAND(NAND(NAND(x,x),NAND(x,x)),
NAND(NAND(y,y),NAND(y,y))),NAND(NAND(NAND(x,x),NAND(x,x)),
NAND(NAND(y,y),NAND(y,y)))), NAND(z,z)), NAND(NAND(NAND(NAND(x,x),
NAND(NAND(y,y), NAND(y,y))), NAND(NAND(x,x),NAND(NAND(y,y),
NAND(y,y)))),NAND(NAND(z,z),NAND(z,z)))),NAND(NAND(NAND(NAND(
NAND(NAND(x,x), NAND(x,x)), NAND((NAND(y,y),NAND(y,y))),NAND(NAND(NAND(x,x),
NAND(x,x)), NAND(NAND(y,y),NAND(y,y)))), NAND(z,z)), NAND(NAND(NAND(NAND(x,x),
NAND(NAND(y,y), NAND(y,y))), NAND(NAND(x,x),NAND(NAND(y,y), NAND(y,y))))),
NAND(NAND(z,z),
NAND(z,z)))))

NOR(NOR(NOR(NOR(NOR(NOR(NOR(x,x), NOR(x,x)),NOR(y,y)),z),
NOR(NOR(NOR(NOR(x,x), NOR(x,x)), NOR(y,y)),z)),
NOR(NOR(NOR(NOR(NOR(x,x), NOR(x,x)), NOR(y,y)), z),
NOR(NOR(NOR(NOR(x,x), NOR(x,x)), NOR(y,y)),z))),

NOR(NOR(NOR(NOR(NOR(x,x), NOR(NOR(y,y), NOR(y,y))),
NOR(z,z)), NOR(NOR(NOR(x,x), NOR(NOR(y,y), NOR(y,y))),NOR(z,z))),
NOR(NOR(NOR(NOR(x,x), NOR(NOR(y,y),NOR(y,y))), NOR(z,z)),
NOR(NOR(NOR(x,x),NOR(NOR(y,y),NOR(y,y))), NOR(z,z)))))

5. The Boolean function describing this Venn diagram is
$f(0,0,0) = 0 \quad f(1,0,0) = 0 \quad f(0,1,0) = 1 \quad f(0,0,1) = 0$
$f(1,1,0) = 0 \quad f(1,0,1) = 1 \quad f(0,1,1) = 0 \quad f(1,1,1) = 0$.
or $f(x,y,z) = x\overline{y}z + \overline{x}y\overline{z}$.

Section 6.3 Exercises

1. Not a well-formed formula. There must be parentheses around a negation.
3. This is a well-formed formula by parts (i) and (ii) of the definition.
5. This is a well-formed formula by parts (i) and (ii) of the definition.
7. This is a well-formed formula by part (ii) of the definition.
9. Not a well-formed formula. Parentheses are needed about $\exists x$.

11. The sentence $(\sim (\exists x) A(x))$ where $A(x)$ is the sentence "The square of x is 2." If the interpretation S is the set of integers, the sentence is true. If S is the set of real numbers, the statement is false.

13. The general form is given below along with a specific example where $A(x)$ means "x is an athlete", $B(x)$ means "x is a baseball player" and $C(x)$ means "x is a millionaire."

$(\forall x)(B(x) \to A(x))$	All baseball players are athletes.
$(\exists x)(C(x) \wedge B(x))$	Some millionaires are baseball players.
$(\exists x)(C(x) \wedge A(x))$	Some millionaires are athletes.

15. The general form is given below along with a specific example where $A(x)$ means "x is a physics teacher", $B(x)$ means "x is a science teacher" and $C(x)$ means "x is a high school teacher."

$(\forall x)(A(x) \to B(x))$	All physics teachers are science teachers.
$(\exists x)(C(x) \wedge (\sim (B(x))))$	Some high school teachers do not teach science.
$(\exists x)(C(x) \wedge (\sim (A(x))))$	Some high school teachers do not teach physics.

17. The general form is given below along with a specific example where $A(x)$ means "x is a sled dog", $B(x)$ means "x is an Alaskan Malamute" and $C(x)$ means "x is a working dog."

$(\exists x)(B(x) \wedge A(x))$	Some Alaskan Malamutes are sled dogs.
$(\forall x)(B(x) \to C(x))$	All Alaskan Malamutes are working dogs.
$(\exists x)(C(x) \wedge A(x))$	Some working dogs are sled dogs.

19. The general form is given below along with a specific example where $A(x)$ means "x is an integer", $B(x)$ means "x is a rational number" and $C(x)$ means "x is an irrational number."

$(\forall x)(A(x) \to B(x))$	All integers are rational numbers.

$(\forall x)(B(x) \to (\sim (C(x))))$ No rational number is irrational.
$(\forall x)(C(x) \to (\sim (A(x))))$ No irrational numbers are integers.

21. $(\forall x)(M(x) \to N(x))$ where $M(x)$ is "x is a man" and $N(x)$ is "x is the nephew of his father's brother."

23. $(\forall x)(O(x) \to (\exists y)(I(y) \wedge B(x,y)))$ where $O(x)$ is "x is greater than 1", $I(y)$ is "y is an integer" and $B(x,y)$ is "y is between x and $2x$."

25. $(\exists x)(A(x) \wedge \sim B(x))$

27. $(\forall x)(\exists y)(\forall z)(A(z) \wedge \sim B(x,y))$

Section 6.4 Supplementary Exercises

1.

(x,y)	f(x,y)
(0, 0)	1
(0, 1)	1
(1, 1)	1
(1, 0)	0

3.

(x,y,z)	f(x,y,z)
(0, 0, 0)	1
(0, 0, 1)	1
(0, 1, 1)	0
(0, 1, 0)	1
(1, 1, 0)	1
(1, 1, 1)	1
(1, 0, 1)	1
(1, 0, 0)	1

5.

(x,y,z)	f(x,y,z)
(0, 0, 0)	0
(0, 0, 1)	0
(0, 1, 1)	1
(0, 1, 0)	0
(1, 1, 0)	0
(1, 1, 1)	1
(1, 0, 1)	1
(1, 0, 0)	0

7.

(x,y,z,w)	f(x,y,z,w)
(0, 0, 0, 0)	0
(0, 0, 0, 1)	0
(0, 0, 1, 1)	1

(x,y,z,w)	$f(x,y,z,w)$
(0, 0, 1, 0)	0
(0, 1, 1, 0)	0
(0, 1, 1, 1)	1
(0, 1, 0, 1)	0
(0, 1, 0, 0)	0
(1, 1, 0, 0)	1
(1, 1, 0, 1)	1
(1, 1, 1, 1)	1
(1, 1, 1, 0)	1
(1, 0, 1, 0)	1
(1, 0, 1, 1)	1
(1, 0, 0, 1)	0
(1, 0, 0, 0)	0

9.

(x,y,z,w)	$f(x,y,z,w)$
(0, 0, 0, 0)	0
(0, 0, 0, 1)	0
(0, 0, 1, 1)	0
(0, 0, 1, 0)	0
(0, 1, 1, 0)	0
(0, 1, 1, 1)	1
(0, 1, 0, 1)	0
(0, 1, 0, 0)	0
(1, 1, 0, 0)	1
(1, 1, 0, 1)	1
(1, 1, 1, 1)	1
(1, 1, 1, 0)	1
(1, 0, 1, 0)	1
(1, 0, 1, 1)	1
(1, 0, 0, 1)	1
(1, 0, 0, 0)	1

11. $f(x,y) = \bar{x} + y = \bar{x}y + \bar{x}\bar{y} + xy$

13. $f(x,y,z) = x + \bar{y} + \bar{z} = xyz + xy\bar{z} + x\bar{y}z + x\bar{y}\bar{z} + \bar{x}y\bar{z} + \bar{x}\bar{y}z + \bar{x}\bar{y}\bar{z}$

15. $f(x,y,z) = xz + yz = xyz + x\bar{y}z + \bar{x}yz$

17. $f(x,y,z,w) = xy + xz + wz = xyzw + xyz\bar{w} + xy\bar{z}w + xy\bar{z}\bar{w} + \bar{x}yzw + \bar{x}\bar{y}zw + x\bar{y}zw + x\bar{y}z\bar{w}$

19. $f(x,y,z,w) = x + yzw = xyzw + xyz\bar{w} + xy\bar{z}w + xy\bar{z}\bar{w} + x\bar{y}zw + x\bar{y}z\bar{w} + x\bar{y}\bar{z}w + x\bar{y}\bar{z}\bar{w} + \bar{x}yzw$

21. Using the Karnaugh map below, we have $\bar{x}y + \bar{x}\bar{y} = \bar{x}(y + \bar{y}) = \bar{x}$, $xy + \bar{x}y = (x + \bar{x})y = y$ so $f(x,y) = \bar{x} + y$.

	y	$\bar{y}$
x	1	0
$\bar{x}$	1	1

23. Using the Karnaugh map below, we have $xyz + xy\bar{z} + x\bar{y}z + x\bar{y}\bar{z} = x(y + \bar{y})(z + \bar{z}) = x$,

$xy\overline{z} + x\overline{y}\overline{z} + \overline{x}y\overline{z} + \overline{x}\overline{y}\overline{z} = (x+\overline{x})(y+\overline{y})\overline{z} = \overline{z}$ and $x\overline{y}z + \overline{x}\overline{y}z = (x+\overline{x})\overline{y}z = \overline{y}z$
so $f(x,y,z) = x + \overline{z} + \overline{y}z$.

	yz	$y\overline{z}$	$\overline{y}\overline{z}$	$\overline{y}z$
x	1	1	1	1
$\overline{x}$	0	1	1	1

25. Using the Karnaugh map below, we have $xyz + \overline{x}yz = (x+\overline{x})yz = yz$ and
$xyz + x\overline{y}z = x(y+\overline{y})z = xz$ so $f(x,y,z) = xz + yz$

	yz	$y\overline{z}$	$\overline{y}\overline{z}$	$\overline{y}z$
x	1	0	0	1
$\overline{x}$	1	0	0	0

27. Using the Karnaugh map below, we have $x\overline{w}yz + xwyz = x(w+\overline{w})yz = xyz$,
$xwyz + \overline{x}wyz + xw\overline{y}z + \overline{x}w\overline{y}z = (x+\overline{x})w(y+\overline{y})z = wz$ and
$xwyz + xwy\overline{z} + x\overline{w}yz + x\overline{w}y\overline{z} = x(w+\overline{w})y(z+\overline{z}) = xy$ so $f(x,y,z,w) = xy + wz + x\overline{y}z$.

	yz	$y\overline{z}$	$\overline{y}\overline{z}$	$\overline{y}z$
xw	1	1	0	1
$\overline{x}w$	1	0	0	1
$\overline{x}\overline{w}$	0	0	0	0
$x\overline{w}$	1	1	0	1

29. Using the Karnaugh map below, we have $xwyz + \overline{x}wyz = (x+\overline{x})wyz = wyz$ and
$xwyz + xwy\overline{z} + xw\overline{y}z + xw\overline{y}\overline{z} + x\overline{w}yz + x\overline{w}y\overline{z} + x\overline{w}\overline{y}z + x\overline{w}\overline{y}\overline{z} = x(w+\overline{w})(y+\overline{y})(z+\overline{z}) = x$
$f(x,y,z,w) = wyz + x$.

	yz	$y\bar{z}$	$\bar{y}\bar{z}$	$\bar{y}z$
xw	1	1	1	1
$\bar{x}w$	1	0	0	0
$\bar{x}\bar{w}$	0	0	0	0
$x\bar{w}$	1	1	1	1

Chapter 7 Graphs and Relations

Section 7.1 Exercises

1.
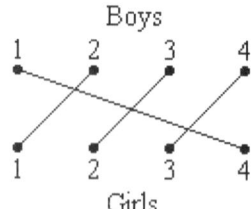

3.
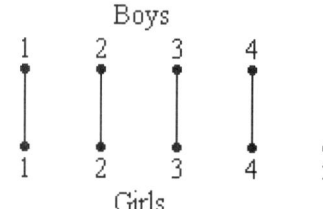

5. Referring to the labeled graph below, a minimal AB-cutset contains two edges, one such being the edges A1 and A3. Two edge-disjoint paths from A to B are (A, 1, 2, B) and (A, 3, 4, B).

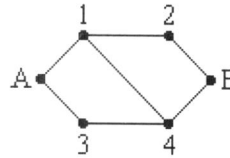

7. Referring to the labeled graph below, a minimal AB-cutset contains three edges, one such being the edges A1, A3, and A5. Three edge-disjoint paths from A to B are (A, 1, 7, 2, B), (A, 3, 8, 4, B) and (A, 5, 7, 6, B).

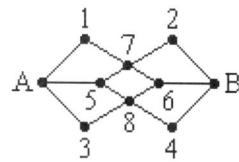

In Exercises 9-19, the desired colorings are given where the colors are represented by the letters A, B, C, D, ... The coloring is applied to the vertices of the associated graph from left to right and top to bottom.

9. A, B, B, C, A, C, A, B
11. A, B, A, B, C, D, C, D
13. A, B, C, D, D, B, A, B

15. A, B, C, D, B, C, D, A

17.
A, B, B, A, A, A   A, C, C, A, A, A   B, A, A, B, B, B
B, C, C, B, B, B   C, A, A, C, C, C   C, B, B, C, C, C
A, B, B, C, C, C   A, B, B, C, A, A   A, B, B, A, C, C
A, B, C, A, C, B

19.
A, B, A, A   A, B, A, C   A, B, C, A   A, B, C, C
A, C, A, A   A, C, A, B   A, C, B, A   A, C, B, B
C, B, A, C   C, A, B, B

Section 7.1 Advanced Exercises

1.

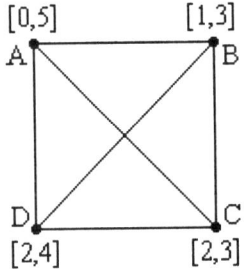

3.

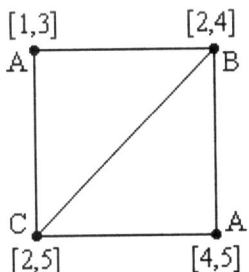

5.

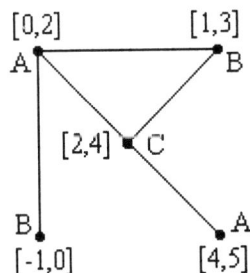

80 Chapter 7 Graphs and Relations

7. Not an interval graph. Labeling three of the vertices leads to an arrangement such as

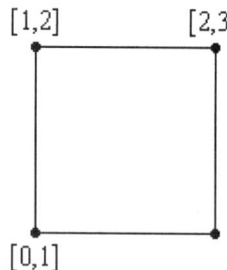

There is no way to label the fourth vertex with an interval since no interval can intersect both [0,1] and [2,3] without intersecting [1,2].

9. This is an interval graph. To see this, label the vertex of degree 3 with the interval [0,5] and the remaining vertices with [0,1], [2,3] and [4,5].

Section 7.2 Exercises

In Exercises 1 through 5 the elements of a maximal antichain of length N are circled in the graph on the left. A set of N disjoint chains is circled on the right.

1. $N = 3$

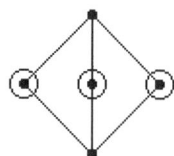 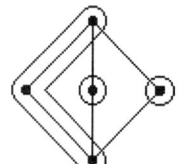

3. $N = 4$

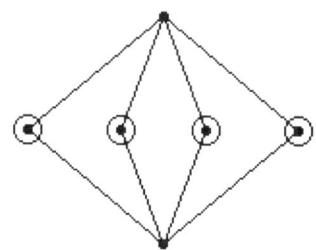 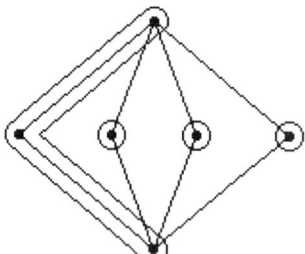

5. $N = 3$

7. This poset is a complemented lattice but is not distributive. To see distributivity fail, label the vertices I, a, b, c, O in order from left to right and top to bottom. Then $a \wedge (b \vee c) = a \wedge I = a$ while $(a \wedge b) \vee (a \wedge c) = O \vee c = c$.

9. This poset is not a lattice. The two elements at the top do not have a join.

11. The complementary pairs of elements of Div(30) are (1,30), (2,15), (3,10), and (5,6).

13. Labeling this lattice as in Exercise 7 above, the element b has two complements, namely a and c.

15. The vertices in the diagram below correspond to the subspaces of the 3-cube as follows. The vertex in the first row is the 3-cube itself. The second row consists of, in order, the seven two-dimensional subspaces generated by the pairs of vectors

{(1,0,0), (0,1,0)}   {(1,0,0), (0,0,1)}   {(1,0,0), (0,1,0)}   {(0,1,0), (0,0,1)}
{(0,1,0), (1,0,1)}   {(0,0,1), (1,1,0)}   {(1,1,0), (1,0,1)}

The third row consists of, in order, the seven one-dimensional subspaces generated by single vectors

(1,0,0) (0,1,0) (0,0,1) (1,1,0) (1,0,1) (0,1,1) (1,1,1).

The vertex in the fourth row represents the zero subspace.

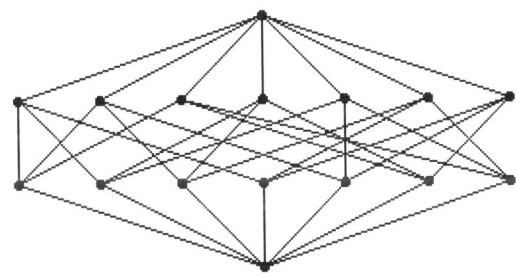

17. The binary expansion of 2 is 10. The sum $(1,0) + (1,0) + (1,0) = (1,0)$ is not the zero vector so 2,2,2 is an unsafe state for player A.

19. The binary expansion of 2 is 10 and that for 3 is 11. The sum $(1,0) + (1,1) + (1,0) + (1,1) = (0,0)$ is the zero vector so 2,3,2,3 is a safe state for player A.

Section 7.2 Advanced Exercises

1. Let Γ be an antichain of the n-cube of maximal length $\binom{n}{\lfloor n/2 \rfloor}$. From the proof of Sperner's Lemma, we know $\sum_{A \in \Gamma} \binom{n}{|A|}^{-1} \leq 1$. Note equality holds if for each $A \in \Gamma$, $|A| = \lfloor n/2 \rfloor$. Further if Γ contains any element A for which $|A| \neq \lfloor n/2 \rfloor$, then $\binom{n}{|A|} < \binom{n}{\lfloor n/2 \rfloor}$ since $\binom{n}{\lfloor n/2 \rfloor}$ is maximal among binomial coefficients $\binom{n}{k}$.

82 Chapter 7 Graphs and Relations

Hence $\binom{n}{|A|}^{-1} > \binom{n}{\lfloor n/2 \rfloor}^{-1}$ and the summation would exceed 1, a contradiction. This proves that an antichain Γ of the n-cube of maximal length must have all its elements of the same cardinality, $\lfloor n/2 \rfloor$. For any k the set of subsets of $\{1,2,...,n\}$ of size k is an antichain of length $\binom{n}{k}$. If n is even, there is only such of maximal length, when $k = n/2$. If n is odd there are two such antichains of maximal length for $k = \frac{n-1}{2}$ and $k = \frac{n+1}{2}$ since $\binom{n}{\lfloor n/2 \rfloor} = \binom{n}{\frac{n-1}{2}} = \binom{n}{n - \frac{n-1}{2}} = \binom{n}{\frac{n+1}{2}}$.

3. Label the vertices of both graphs a, b, c, d, and e, in order, from top to bottom and left to right. In the first graph, we have $b \wedge (c \vee d) = b \wedge a = b$ while $(b \wedge c) \vee (b \wedge d) = e \vee d = d$ and so the first lattice is not distributive.

In the second graph, $c \wedge (b \vee d) = c \wedge a = c$ while $(c \wedge b) \vee (c \wedge d) = e \vee e = e$ and so the second lattice is not distributive.

5. Let V_1, V_2, and V_3 be subspaces of the n-cube with $V_1 \subset V_3$. Suppose $x \in V_1 + (V_2 \cap V_3)$ then $x = v + w$ where $v \in V_1$ and $w \in V_2 \cap V_3$. Since $w \in V_2$, it follows that $x = v + w \in V_1 + V_2$. Also note that since $V_1 \subset V_3$ both v and w are in V_3 so $x = v + w \in V_3$. Hence $x \in (V_1 + V_2) \cap V_3$ and $V_1 + (V_2 \cap V_3) \subseteq (V_1 + V_2) \cap V_3$. Conversely, suppose $x \in (V_1 + V_2) \cap V_3$ then $x \in V_3$ and $x = v + w$ where $v \in V_1$ and $w \in V_2$. Since $V_1 \subset V_3$, both x and v are in V_3 and therefore $v + x = v + (v + w) = (v + v) + w = 0 + w = w \in V_3$. Thus $w \in V_2 \cap V_3$ and so $x \in V_1 + (V_2 \cap V_3)$. This gives $(V_1 + V_2) \cap V_3 \subseteq V_1 + (V_2 \cap V_3)$ and so $V_1 + (V_2 \cap V_3) = (V_1 + V_2) \cap V_3$ and the n-cube is modular.

Section 7.3 Supplementary Exercises

1. No. Boy #2 would have to marry the Girl #2, forcing Boy #3 to marry Girl #6. This leaves Boy #5 no to marry that he knows.

3. A 4-coloring is given by A, B, C, A, B, C, D where the colors are applied to the vertices from left to right and top to bottom.

5. The elements of a maximal antichain of length 2 are circled in the graph on the left. A set of 2 disjoint chains is circled on the right.

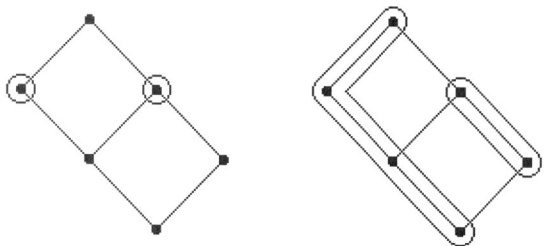

Section 7.3 Advanced Exercises

1. $P_G(k) = k(k-1)(k-2)(k-3)$

3. $P_G(k) = k(k-1)^4$

5. Label the nodes of the tree w, x, y, z in order from left to right and top to bottom. The corresponding poset is listed below where (a,b) means $a \leq b$.
$(w,w), (x,x), (y,y), (z,z), (x,w), (y,x), (y,w), (z,x), (z,w)$.

7. Label the nodes of the tree $r, s, t, u, v, w, x, y, z$ in order from left to right and top to bottom. The corresponding poset is listed below where (a,b) means $a \leq b$.
$(r,r), (s,s), (t,t), (u,u), (v,v), (w,w), (x,x), (y,y), (z,z), (s,r), (t,r), (u,r), (v,r),$
$(w,r), (x,r), (y,r), (z,r), (v,s), (w,s), (x,t), (y,u), (z,u)$.

Chapter 8 Algorithms

Section 8.1 Exercises

1. The steps for the Mergesort are
    (1, 4)(3, 7)(2, 5)(8, 9)
    (1, 3, 4, 7)(2, 5, 8, 9)
    (1, 2, 3, 4, 5, 7, 8, 9)

    The steps for the Quicksort are
    (2, 1, 3)(7, 4, 5, 9, 8)
    (1)(2, 3)(7, 4, 5, 9, 8)
    (1)(2)(3)(7, 4, 5, 9, 8)
    (1)(2)(3)(5, 4)(7, 9, 8)
    (1)(2)(3)(4)(5)(7, 9, 8)
    (1)(2)(3)(4)(5)(7)(9, 8)
    (1)(2)(3)(4)(5)(7)(8)(9)

3. The steps for the Mergesort are
    (1, 2)(4, 5)(3, 9)(6, 7)
    (1, 2, 4, 5)(3, 6, 7, 9)
    (1, 2, 3, 4, 5, 6, 7, 9)

    The steps for the Quicksort are
    (1)(2, 5, 4, 9, 3, 7, 6)
    (1)(2)(5, 4, 9, 3, 7, 6)
    (1)(2)(3, 4)(9, 5, 7, 6)
    (1)(2)(3)(4)(5, 7, 6)(9)
    (1)(2)(3)(4)(5)(7, 6)(9)
    (1)(2)(3)(4)(5)(6)(7)(9)

5. The steps for the Mergesort are
    (7, 11)(13, 19)(1, 18)(5, 12)(4, 16)(3, 17)(2, 15)(10, 14)
    (7, 11, 13, 19)(1, 5, 12, 18)(3, 4, 16, 17)(2, 10, 14, 15)
    (1, 5, 7, 11, 12, 13, 18, 19)(2, 3, 4, 10, 14, 15, 16, 17)
    (1, 2, 3, 4, 5, 7, 10, 11, 12, 13, 14, 15, 16, 17, 18, 19)

    The steps for the Quicksort are
    (2, 3, 4, 5, 1)(18, 19, 12, 13, 16, 17, 11, 15, 7, 14, 10)
    (1)(3, 4, 5, 2)(18, 19, 12, 13, 16, 17, 11, 15, 7, 14, 10)
    (1)(2)(4, 5, 3)(18, 19, 12, 13, 16, 17, 11, 15, 7, 14, 10)
    (1)(2)(3)(5, 4)(18, 19, 12, 13, 16, 17, 11, 15, 7, 14, 10)
    (1)(2)(3)(4)(5)(18, 19, 12, 13, 16, 17, 11, 15, 7, 14, 10)
    (1)(2)(3)(4)(5)(10, 14, 12, 13, 16, 17, 11, 15, 7)(19, 18)
    (1)(2)(3)(4)(5)(7)(14, 12, 13, 16, 17, 11, 15, 10)(19, 18)
    (1)(2)(3)(4)(5)(7)(10, 12, 13, 11)(17, 16, 15, 14)(19, 18)
    (1)(2)(3)(4)(5)(7)(10)(12, 13, 11)(17, 16, 15, 14)(19, 18)
    (1)(2)(3)(4)(5)(7)(10)(11)(13, 12)(17, 16, 15, 14)(19, 18)
    (1)(2)(3)(4)(5)(7)(10)(11)(12)(13)(17, 16, 15, 14)(19, 18)
    (1)(2)(3)(4)(5)(7)(10)(11)(12)(13)(14, 16, 15)(17)(19, 18)
    (1)(2)(3)(4)(5)(7)(10)(11)(12)(13)(14)(16, 15)(17)(19, 18)
    (1)(2)(3)(4)(5)(7)(10)(11)(12)(13)(14)(15)(16)(17)(19, 18)
    (1)(2)(3)(4)(5)(7)(10)(11)(12)(13)(14)(15)(16)(17)(18)(19)

7. $i=1 \quad j=8 \quad k=4 \quad a(k)=9<37$
   $i=5 \quad j=8 \quad k=6 \quad a(k)=50>37$
   $i=5 \quad j=5 \quad k=5 \quad a(k)=37$.

9. $i=1 \quad j=16 \quad k=8 \quad a(k)=40>37$
   $i=1 \quad j=7 \quad k=4 \quad a(k)=20<37$
   $i=5 \quad j=7 \quad k=6 \quad a(k)=31<37$
   $i=7 \quad j=7 \quad k=7 \quad a(k)=37$.

11-13. The chart below lists the lex, revlex, grlex, and grevlex orderings of the ordered pairs of elements from the set {1, 2, 3, 4}. All orderings are listed from smallest element to largest element.

lex	grlex
(1, 1)	(1, 1)
(1, 2)	(1, 2)
(1, 3)	(2, 1)
(1, 4)	(1, 3)
(2, 1)	(2, 2)
(2, 2)	(3, 1)
(2, 3)	(1, 4)
(2, 4)	(2, 3)
(3, 1)	(3, 2)
(3, 2)	(4, 1)
(3, 3)	(2, 4)
(3, 4)	(3, 3)
(4, 1)	(4, 2)
(4, 2)	(3, 4)
(4, 3)	(4, 3)
(4, 4)	(4, 4)

15-17. The chart below lists the lex, revlex, grlex, and grevlex orderings of the three-element subsets of the set {1, 2, 3, 4, 5, 6}. All orderings are listed from smallest element to largest element.

lex	grlex
{1, 2, 3}	{1, 2, 3}
{1, 2, 4}	{1, 2, 4}
{1, 2, 5}	{1, 2, 5}
{1, 2, 6}	{1, 3, 4}
{1, 3, 4}	{1, 2, 6}
{1, 3, 5}	{1, 3, 5}
{1, 3, 6}	{2, 3, 4}
{1, 4, 5}	{1, 3, 6}
{1, 4, 6}	{1, 4, 5}
{1, 5, 6}	{2, 3, 5}
{2, 3, 4}	{1, 4, 6}
{2, 3, 5}	{2, 3, 6}
{2, 3, 6}	{2, 4, 5}
{2, 4, 5}	{1, 5, 6}
{2, 4, 6}	{2, 4, 6}
{2, 5, 6}	{3, 4, 5}
{3, 4, 5}	{2, 5, 6}
{3, 4, 6}	{3, 4, 6}
{3, 5, 6}	{3, 5, 6}
{4, 5, 6}	{4, 5, 6}

Section 8.1 Advanced Exercises

1. Label the three disks 1, 2 and 3 with 1 being the largest and 3 the smallest. When moving through the Standard Gray Code, a change in coordinate i indicates to move disk i. Below we walk through the code describing the corresponding move in the game.

(0, 0, 0): Start of game, disks are stacked in decreasing size
(0, 0, 1): Move disk 3 to one of the empty pegs
(0, 1, 1): Move disk 2 to the only empty peg
(0, 1, 0): Place disk 3 on top of disk 2
(1, 1, 0): Move disk 1 to the peg just vacated by disk 3
(1, 1, 1): Move disk 3 to the peg just vacated by disk 1
(1, 0, 1): Place disk 2 on top of disk 1
(1, 0, 0): Place disk 3 on top of disk 2.

Section 8.2 Exercises

1. Both Prim and Kruskal lead to the minimum-weight spanning tree below.

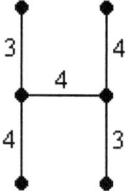

3. Both Prim and Kruskal lead to the minimum-weight spanning tree below.

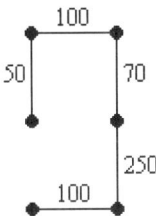

5. Both Prim and Kruskal lead to the minimum-weight spanning tree below.

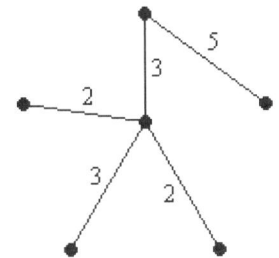

7.

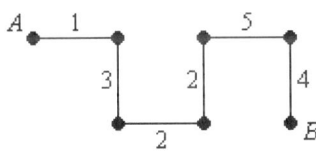

9.

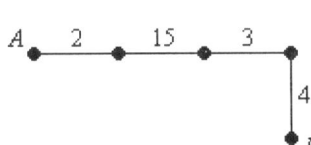

11. Applying Kruskal's algorithm to this graph will always require the selection of the two edges of weight 1, in some order, followed by the two edges of weight 2 in some order. The last edge to complete the spanning tree may be either edge of weight 3. Choosing the lower such edge gives the result below.

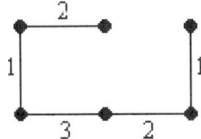

13. The minimal cut is indicated by the dashed lines below. The maximal flow from A to B is 9, the sum of the cut edge values.

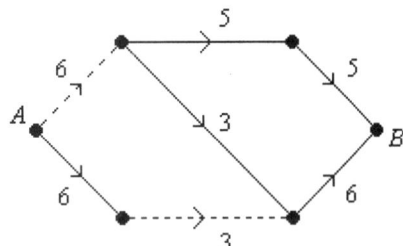

15. The minimal cut is indicated by the dashed lines below. The maximal flow from A to B is 4, the sum of the cut edge values.

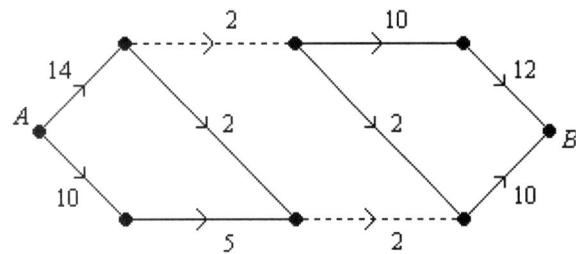

Section 8.3 Exercises

1. Since $0 \le x^2 + 2$ for all x and $x^2 + 2 \le 2x^2$ for $2 \le x^2$ or $\sqrt{2} \le x$ it follows that $x^2 + 2 = O(x^2)$.

3. Since $0 \le x^3 + 2$ for all $0 \le x$ and $x^3 + 2 \le 2x^3$ for $2 \le x^3$ or $\sqrt[3]{2} \le x$ it follows that $x^3 + 2 = O(x^3)$.

5. First note $x \le 2^x$ for $x \ge 1$. This can be verified by graphing the function or through calculus by noting that $f(x) = 2^x - x$ satisfies $f(1) = 1$ and is increasing for $x \ge 1$ since $f'(x) = 2^x(\ln 2) - 1 > 0$. It follows that $\lg(x) \le \lg(2^x) = x$ so $x\lg(x) \le x^2$ for $x \ge 1$. Since we also have $0 \le x\lg(x)$ for $x \ge 1$ it follows that $x\lg(x) = O(x^2)$.

7. For $x \ge 500$ the term $2x - 1000$ is positive so $0 \le x^2 \le x^2 + 2x - 1000$ which shows $x^2 + 2x - 1000 = \Omega(x^2)$.

9. If $x \ge 2^{1001}$ then $\lg(x) - 1000 \ge 1$ and $x\lg(x) - 1000x + 500 \ge x\lg(x) - 1000x \ge x \ge 0$. Thus $x\lg(x) - 1000x + 500 = \Omega(x)$.

11. If $x \ge 100$ then $0 \le x \le x + 100 \le x + x = 2x$ so $x + 100 = \Theta(x)$.

13. For any x, $\frac{1}{2}x^2 \le x^2 - x + 1$ since $x^2 - x + 1 - \frac{1}{2}x^2 = \frac{1}{2}x^2 - x + 1 = \frac{1}{2}\left[(x-1)^2 + 1\right] \ge 0$.

If $x \ge 1$ then $x^2 - (x-1) \le x^2$. Thus for $x \ge 1$, $0 \le \frac{1}{2}x^2 \le x^2 - x + 1 \le x^2$ and $x^2 - x + 1 = \Theta(x^2)$.

15. Since $\lg x$ and $x\lg x$ are increasing functions and $\lg 32 = 5$ so $32\lg 32 = 160 > 100$, it follows that for $x \geq 32$, $0 \leq x\lg x \leq x\lg x + x + 100 \leq x\lg x + x\lg x + x\lg x = 3x\lg x$ and so $x\lg x + x + 100 = \Theta(x\lg x)$.

17. $\dfrac{1}{10^9} \sum_{k=1}^{30} 2^k = 2.15$ seconds.

19. $\dfrac{1}{10^9} \sum_{k=1}^{50} 2^k = 26.06$ days.

21. $\dfrac{1}{10^9} \sum_{k=1}^{1000000} k = \dfrac{(1000000)(1000001)}{2 \cdot 10^9} = 500$ seconds or 8.33 minutes.

23. $\dfrac{1}{10^9} \sum_{k=1}^{10000} k^2 = \dfrac{(10000)(10001)(20001)}{6 \cdot 10^9} = 333.38$ seconds or about five and a half minutes.

25. By Bernoulli's formula $\sum_{k=1}^{n} k^{10} = \dfrac{(n+1)^{11}}{11} - \dfrac{(n+1)^{10}}{2} + \dfrac{5}{6}(n+1)^9 - (n+1)^7 + (n+1)^5 - \dfrac{(n+1)^3}{2} + \dfrac{5}{66}(n+1)$.

Applying this formula to $\dfrac{1}{10^9} \sum_{k=1}^{1000} k^{10}$ gives a total time of approximately 2.9×10^{15} years.

27. Stirling: $8! = \sqrt{16\pi}\left(\dfrac{8}{e}\right)^8 = 39902.3955$

   Series: $8! = \sqrt{16\pi}\left(\dfrac{8}{e}\right)^8\left(1 + \dfrac{1}{12 \cdot 9} + \dfrac{1}{288 \cdot 9^2}\right) = 40273.5726$

A relative error of about 0.92%.

29. Stirling: $20! = \sqrt{40\pi}\left(\dfrac{20}{e}\right)^{20} = 2422786846761133393.6834$

   Series: $20! = \sqrt{40\pi}\left(\dfrac{20}{e}\right)^{20}\left(1 + \dfrac{1}{12 \cdot 21} + \dfrac{1}{288 \cdot 21^2}\right) = 2432420156140183314.6634$

A relative error of about 0.4%.

Section 8.3 Advanced Exercises

1. No. While it is true that $0 \leq 2^x \leq 3^x$ for $x \geq 0$, no constant D can be found with $3^x \leq D \, 2^x$ since $\dfrac{3^x}{2^x}$ grows infinitely large with x and so cannot be bounded.

3. No. If $\lg \lg x = \Theta(\lg x)$ then there would be constants C and $x0$ such that $0 \leq C\lg x \leq \lg\lg x$ or $\dfrac{\lg x}{\lg\lg x} \leq \dfrac{1}{C}$ for $x \geq x0$. However $\lim_{x \to \infty} \dfrac{\lg x}{\lg\lg x} = \lim_{x \to \infty} \dfrac{1/(x\ln 2)}{(1/\lg x)(1/(x\ln 2))} = \lim_{x \to \infty} \lg x = \infty$ and so cannot be bounded.

5. Suppose the common degree of f(X) and g(X) is n. Then $\lim_{X \to \infty} \frac{f(X)}{X^n} = f_n$ the highest degree coefficient of $f(X)$. Choose $\varepsilon < f_n$. By the definition of the limit, there is a value X_1 such that $f_n - \varepsilon < \frac{f(X)}{X^n} < f_n + \varepsilon$, i.e. $C_1 X^n < f(X) < D_1 X^n$ for constants C_1 and D_1 and $X > X_1$. Similarly, there are constants X_2, C_2 and D_2 such that $C_2 X^n < g(X) < D_2 X^n$ for $X > X_2$. Then for $X > \max\{X_1, X_2\}$ we have $\frac{C_1}{D_2} g(X) < \frac{C_1}{D_2} D_2 X^n = C_1 X^n < f(X) < D_1 X^n = \frac{D_1}{C_2} C_2 X^n < \frac{D_1}{C_2} g(X)$. Since $f(X)$ is between constant multiples of $g(X)$ for X suitably large, $f(X) = \Theta(g(X))$.

Section 8.4 Supplementary Exercises

1. Both the Kruskal and Prim algorithms could lead to either minimum weight spanning tree below.

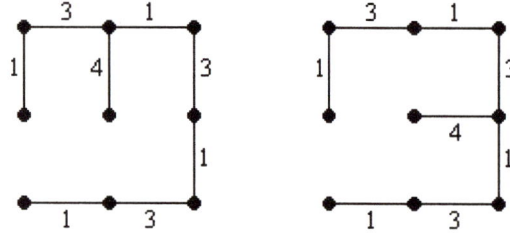

3. The shortest path from A to B is shown below.

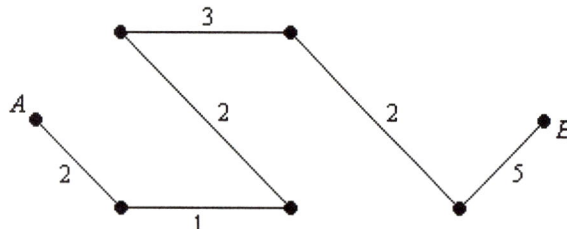

5. The minimal cut is indicated by the dashed lines below. The maximal flow from A to B is 6, the sum of the cut edge values.

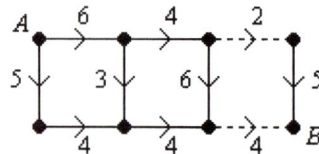

Chapter 9 Combinatorics

Section 9.1 Exercises

Unless otherwise noted, the solutions to exercises 1-9 were found primarily by inspection.

1. $a_n = a_{n-1} + 1$, $a_0 = 1$

3. $a_n = a_{n-2}$, $a_0 = 1$, $a_1 = 3$

5. $a_n = 3a_{n-1} - 3a_{n-2} + a_{n-3}$, $a_0 = 1$, $a_1 = 4$, $a_2 = 9$

By inspection, the general term of the given sequence is $a_n = (n+1)^2$. To express a_n as a linear recursion, assume $a_n = k_1 a_{n-1} + k_2 a_{n-2} + k_3 a_{n-3}$. Substituting the first four terms of the sequence into this relation gives the equation

$$16 = 9k_1 + 4k_2 + k_3$$

Similarly, the 2nd through 4th and 3rd through 5th terms yield

$$25 = 16k_1 + 9k_2 + 4k_3$$
$$36 = 25k_1 + 16k_2 + 9k_3$$

Solving the resulting system of equations gives $k_1 = 3$, $k_2 = -3$, $k_3 = 1$.

7. $a_n = a_{n-1} + a_{n-2}$, $a_0 = 2$, $a_1 = 2$

9. $a_n = a_{n-1} + a_{n-2} + a_{n-3}$, $a_0 = 1$, $a_1 = -1$, $a_2 = 0$

11. Using the geometric series $\dfrac{1}{1-t} = \sum\limits_{k=0}^{\infty} t^k$ (i.e. the inverse binomial series with $n = 1$) and substituting $t = x^2$ gives $\dfrac{1}{1-x^2} = \sum\limits_{k=0}^{\infty} x^{2k}$.

13. Using the geometric series $\dfrac{1}{1-t} = \sum\limits_{k=0}^{\infty} t^k$ and substituting $t = x^3$ gives $\dfrac{1}{1-x^3} = \sum\limits_{k=0}^{\infty} x^{3k}$.

15. The inverse of $(1-x)^3$ is given immediately by the inverse binomial series with $n = 3$:

$$\frac{1}{(1-x)^3} = \sum_{k=0}^{\infty} \binom{k+2}{k} x^k = \sum_{k=0}^{\infty} \frac{(k+1)(k+2)}{2} x^k .$$

17. We apply the inverse generating algorithm using $a_0 = a_1 = a_2 = 1$ and $a_k = 0$ for $k > 2$. The first three equations of the algorithm are $b_0 = 1, b_1 + b_0 = 0$, and $b_2 + b_1 + b_0 = 0$ with the general equation for $k > 2$ being $b_k + b_{k-1} + b_{k-1} = 0$. Solving for the first three coefficients gives $b_0 = 1, b_1 = -1$ and $b_2 = 0$. Using these values and the general equation, we find the next three coefficients must satisfy $b_3 = 1, b_4 + b_3 = 0$, and $b_5 + b_4 + b_3 = 0$. This system of three equations in three unknowns is identical to the system involving the

first three coefficients. Thus $b_3 = 1, b_4 = -1$ and $b_5 = 0$. This pattern must repeat giving the inverse power series

$$\frac{1}{1+x+x^2} = 1 - x + x^3 - x^4 + x^6 - x^7 + \ldots = \sum_{k=0}^{\infty} b_k x^k \text{ where } b_{3n} = 1, b_{3n+1} = -1, b_{3n+2} = 0 \text{ for } n > 0.$$

19. Applying the inverse generating algorithm with $a_0 = 2, a_1 = 0, a_2 = -1$ and $a_k = 0$ for $k > 2$ gives $2b_0 = 1, 2b_1 = 0$ and $2b_k - b_{k-2} = 0$ for $k \geq 2$. It follows that $b_0 = 1/2, b_1 = 0$ and $b_k = (1/2)b_{k-2}$ for $k \geq 2$. From the recursion formula and the fact $b_0 = 1/2$, it follows that $b_{2k} = \frac{1}{2^{k+1}}$ while the odd terms must all equal 0. The inverse power series is then $\sum_{k=0}^{\infty} \frac{x^{2k}}{2^{k+1}}$. Note this same result could be found by replacing x with $\frac{x}{\sqrt{2}}$ in the result of Exercise 11 and multiplying by 2.

21. The generating function for a_n is given by

$$f(x) = 1 + 2x + 4x^2 + 8x^3 + \ldots = 1 + (2x) + (2x)^2 + (2x)^3 + \ldots = \frac{1}{1-2x} = \sum_{n=0}^{\infty} 2^n x^n.$$

Thus $a_n = 2^n$.

23. $f(x) = 3 + 5x + 9x^2 + 17x^3 + \ldots = (2 + 4x + 8x^2 + 16x^3 + \ldots) + (1 + x + x^2 + x^3 + \ldots)$

Noting the first series is two times the series of Exercise 21 while the second is the geometric series, we find the generating function (and its partial fraction expansion) to be

$$f(x) = \frac{2}{1-2x} + \frac{1}{1-x} = 2\sum_{n=0}^{\infty} 2^n x^n + \sum_{n=0}^{\infty} x^n = \sum_{n=0}^{\infty} (2^{n+1} + 1)x^n \text{ so } a_n = 2^{n+1} + 1.$$

25. $f(x) = 1 + x + x^2 + 2x^3 + 2x^4 + 2x^5 + 3x^6 + 3x^7 + 3x^8 \ldots$

$$= (1 + x + x^2) + 2x^3(1 + x + x^2) + 3x^6(1 + x + x^2) + \ldots$$

$$= (1 + x + x^2)(1 + 2x^3 + 3x^6 + \ldots) = \frac{1 + x + x^2}{(1-x^3)^2} = \frac{1}{(1-x)(1-x^3)}.$$

The identity $1 + 2x^3 + 3x^6 + \ldots = \frac{1}{(1-x^3)^2}$ used above is determined by first applying the inverse binomial series with $n = 2$ giving $\frac{1}{(1-t)^2} = \sum_{k=0}^{\infty} \binom{k+1}{k} t^k = \sum_{k=0}^{\infty} (k+1)t^k = 1 + 2t + 3t^2 + \ldots$ and then substituting $t = x^3$.

Expanding $\frac{1}{(1-x)(1-x^3)}$ in partial fractions gives $\frac{1/3}{(1-x)^2} + \frac{1/3}{1-x} + \frac{(1/3)(1+x)}{1+x+x^2}$. Expanding each term in a series gives $\sum_{n=0}^{\infty} \frac{1}{3}(n+1)x^n + \sum_{n=0}^{\infty} \frac{1}{3} x^n + \sum_{n=0}^{\infty} \frac{1}{3}\left(\frac{1}{r^n(1-r)} + \frac{1}{\overline{r}^n(1-\overline{r})}\right) x^n$ where $r = \frac{-1+i\sqrt{3}}{2}$, the last series arising from the observation that $\frac{1+x}{1+x+x^2} = \frac{1}{1-r}\left(\frac{1}{1-x/r}\right) + \frac{1}{1-\overline{r}}\left(\frac{1}{1-x/\overline{r}}\right)$. Thus

$$a_n = \frac{1}{3}(n+1) + \frac{1}{3} + \frac{1}{3}\left(\frac{1}{r^n(1-r)} + \frac{1}{\overline{r}^n(1-\overline{r})}\right).$$

27. Here $a=2$, $b=3$ and $\log_b(a) \approx 0.631$. Let $\varepsilon = \log_3 2 - 0.5 \approx 0.131$ then $f(n) = \sqrt{n} = O(\sqrt{n}) = O(n^{\log_3 2 - \varepsilon})$. By the Master Theorem $T(n) = \Theta(n^{\log_3 2})$.

29. Here $a=4$, $b=2$ and $\log_4(2) = 2$. Since $f(n) = n^2 = \Theta(n^2)$ it follows from the Master Theorem that $T(n) = \Theta(n^2 \lg n)$.

Section 9.1 Advanced Exercises

1. The first thirty terms of Conway's sequence are 1, 1, 2, 2, 3, 4, 4, 4, 5, 6, 7, 7, 8, 8, 8, 8, 9, 10, 11, 12, 12, 13, 14, 14, 15, 15, 15, 16, 16, 16.

3. The first thirty terms of Hofstadter's sequence are 1, 1, 2, 3, 3, 4, 5, 5, 6, 6, 6, 8, 8, 8, 10, 9, 10, 11, 11, 12, 12, 12, 12, 16, 14, 14, 16, 16, 16, 16.

1. The first six derivatives of $\tan(x)$ are $\sec^2(x)$, $2\sec^2(x)\tan(x)$, $4\sec^2(x)\tan^2(x) + 2\sec^4(x)$, $16\sec^4(x)\tan(x) + 8\sec^2(x)\tan^3(x)$, $16\sec^2(x)\tan^4(x) + 88\sec^4(x)\tan^2(x) + 16\sec^6(x)$ and $32\sec^2(x)\tan^5(x) + 416\sec^4(x)\tan^3(x) + 272\sec^6(x)\tan(x)$. Evaluating at $x=0$ gives the tangent numbers 1, 0, 2, 0, 16, 0.

The first six derivatives of $\sec(x)$ are $\sec(x)\tan(x)$, $\sec^3(x) + \sec(x)\tan^2(x)$, $5\sec^3(x)\tan(x) + \sec(x)\tan^3(x)$, $18\sec^3(x)\tan^2(x) + 5\sec^5(x) + \sec(x)\tan^4(x)$, $58\sec^3(x)\tan^3(x) + 61\sec^5(x)\tan(x) + \sec(x)\tan^5(x)$, and $179\sec^3(x)\tan^4(x) + 479\sec^5(x)\tan^2(x) + 61\sec^7(x) + \sec(x)\tan^6(x)$. Evaluating at $x=0$ gives the secant numbers 0, 1, 0, 5, 0, 61.

Section 9.2 Exercises

In exercises 1-4, we use the fact that there are 36 possible outcomes from a roll of two dice: (1,1), (1,2), ... (6,5), and (6,6), each one equally likely.

1. $P(X=6) = P(\{(1,5),(2,4),(3,3),(4,2),(5,1)\}) = \dfrac{5}{36}$

3. $P(X \leq 4) = P(\{(1,1),(1,2),(2,1),(1,3),(2,2),(3,1)\}) = \dfrac{6}{36} = \dfrac{1}{6}$

5. $P(X=3) = \binom{5}{3}\dfrac{1}{2^5} = \dfrac{5}{16}$

7. $P(X=5) = \binom{7}{5}\dfrac{1}{2^7} = \dfrac{21}{128}$

9. $P(X \leq 2) = \binom{5}{0}\frac{1}{2^5} + \binom{5}{1}\frac{1}{2^5} + \binom{5}{2}\frac{1}{2^5} = \frac{16}{32} = \frac{1}{2}$

11. $P(2 \leq X \leq 4) = \binom{7}{2}\frac{1}{2^7} + \binom{7}{3}\frac{1}{2^7} + \binom{7}{4}\frac{1}{2^7} = \frac{91}{128}$

13. $P(3 \leq X \leq 5) = \binom{8}{3}\frac{1}{2^8} + \binom{8}{4}\frac{1}{2^8} + \binom{8}{5}\frac{1}{2^8} = \frac{182}{256} = \frac{91}{128}$

15. $P(X = 3) = \binom{5}{3} 0.3^3 0.7^2 = 0.1323$

17. $P(X \leq 2) = \binom{7}{0} 0.3^0 0.7^7 + \binom{7}{1} 0.3^1 0.7^6 + \binom{7}{2} 0.3^2 0.7^5 = 0.6470695$

19. $P(X \leq 2) = \binom{8}{0} 0.3^0 0.7^8 + \binom{8}{1} 0.3^1 0.7^7 + \binom{8}{2} 0.3^2 0.7^6 = 0.55177381$

21. $P(X \geq 6) = \binom{8}{6} 0.3^6 0.7^2 + \binom{8}{7} 0.3^7 0.7^1 + \binom{8}{8} 0.3^8 0.7^0 = 0.01129221$

23. $P(A \text{ succeeds}) = 1 - P(A \text{ fails}) = 1 - \left(\frac{5}{6}\right)^8 = 0.76743$

$P(B \text{ succeeds}) = 1 - P(B \text{ fails}) = 1 - \left(\frac{5}{6}\right)^{16} - \binom{16}{1}\frac{1}{6}\left(\frac{5}{6}\right)^{15} = 0.77283$

In exercises 24-35, z-scores are computed based on a mean $\mu = np = 100(1/2) = 50$ and standard deviation $\sigma = \sqrt{npq} = \sqrt{100(1/2)(1/2)} = 5$. After a transformation to z-scores, normal probability tables are used to determine the desired probability.

25. $P(50 \leq X \leq 65) = P(\frac{50-50}{5} \leq z \leq \frac{65-50}{5}) = P(0 \leq z \leq 3) = 0.49865$

27. $P(42 \leq X \leq 52) = P(\frac{42-50}{5} \leq z \leq \frac{52-50}{5}) = P(-1.6 \leq z \leq 0.4) = 0.60062$

29. $P(X \leq 40) = P(z \leq \frac{40-50}{5}) = P(z \leq -2) = 0.02275$

31. $P(X \leq 43) = P(z \leq \frac{43-50}{5}) = P(z \leq -1.4) = 0.08076$

33. $P(X \geq 55) = P(z \geq \frac{55-50}{5}) = P(z \geq 1) = 0.15866$

35. $P(X \geq 63) = P(z \geq \frac{63-50}{5}) = P(z \geq 2.6) = 0.00466$

Section 9.2 Advanced Exercises

1. Let $N = |S|$. Then $P(B_m - B_{m+1}) = \frac{|B_m - B_{m+1}|}{N}$ and by the Binomial Inversion Formula

$$P(B_m - B_{m+1}) = \left(\sum \frac{|A_{i_1} \cap ... \cap A_{i_m}|}{N} - \sum \binom{m+1}{m} \frac{|A_{i_1} \cap ... \cap A_{i_{m+1}}|}{N} + ... + (-1)^{n-m} \binom{n}{m} \frac{|A_1 \cap ... \cap A_n|}{N} \right)$$

where the summations are taken over all possible intersections of m sets, $(m+1)$ sets, etc. Now note that for any k there are $\binom{n}{k}$ intersections of the form $A_{i_1} \cap A_{i_2} \cap ... \cap A_{i_k}$ and that

$\frac{|A_{i_1} \cap A_{i_2} \cap ... \cap A_{i_k}|}{N} = P(A_{i_1} \cap A_{i_2} \cap ... \cap A_{i_k}) = \frac{1}{k+1}$ so the summations above are really sums of constants not depending on the summation indices.

Thus, for example, $\sum \frac{|A_{i_1} \cap ... \cap A_{i_m}|}{N} = \sum \frac{1}{m+1} = \binom{n}{m} \frac{1}{m+1}$ and the entire probability expression simplifies

to $P(B_m - B_{m+1}) = \binom{n}{m} \frac{1}{m+1} - \binom{n}{m+1}\binom{m+1}{m} \frac{1}{m+2} + \binom{n}{m+2}\binom{m+2}{m} \frac{1}{m+3} - ... (-1)^{n-m} \binom{n}{m} \frac{1}{n+1}$

$= \sum_{k=0}^{n-m} (-1)^k \binom{n}{m+k}\binom{m+k}{m} \frac{1}{m+k+1}$

By the Trinomial Revision Identity of Section 3.3, $\binom{n}{m+k}\binom{m+k}{m} = \binom{n}{m}\binom{n-m}{k}$. With this identity and the Inverse Expansion Identity from Section 3.3, we have

$P(B_m - B_{m+1}) = \sum_{k=0}^{n-m} (-1)^k \binom{n}{m+k}\binom{m+k}{m} \frac{1}{m+k+1} = \sum_{k=0}^{n-m} (-1)^k \binom{n}{m}\binom{n-m}{k} \frac{1}{m+k+1}$

$= \frac{1}{n+1}\binom{n}{m} \sum_{k=0}^{n-m} (-1)^k \binom{n-m}{k} \frac{n+1}{m+k+1} = \frac{1}{n+1}\binom{n}{m}\binom{n}{m}^{-1} = \frac{1}{n+1}$.

Section 9.3 Exercises

1. $(1, 2)$ 3. $(2, 3)$ 5. $(1, 3, 2)$

7. $(1, 4, 3, 2)$ 9. $(1, 4, 3)$

11. The permutations in exercises 1, 2, 3, 6 and 7 are odd permutations. The permutations in exercises 4, 5, 8, 9 and 10 are even permutations.

13.   e   (1, 2)(3, 4)   (1, 3)(2, 4)   (1, 4)(2, 3)
      (2, 3, 4)   (2, 4, 3)   (1, 3, 4)   (1, 4, 3)
      (1, 2, 4)   (1, 4, 2)   (1, 2, 3)   (1, 3, 2)

15. Label the vertices of hexagon with the numbers 1-6 in a clockwise direction. The elements of D_6 are:
the clockwise rotations:
   e   (1, 2, 3, 4, 5, 6)   (1, 3, 5)(2, 4, 6)   (1, 4)(2,5)(3,6)   (1, 5, 3)(2, 6, 4)   (1, 6, 5, 4, 3, 2)
the reflections across a line through two opposite vertices:
   (2, 6)(3, 5)   (1, 3)(4, 6)   (1, 5)(2, 4)
and the reflections across a line through two opposite sides:
   (1, 2)(3, 6)(4, 5)   (2, 3)(1, 4)(5, 6)   (1, 6)(2, 5)(3, 4) .

17. The number of inequivalent configurations of six keys on a ring is the index, $(S_6 : D_6)$, of the dihedral group D_6 in the symmetric group S_6. By Lagrange's Theorem $(S_6 : D_6) = |S_6|/|D_6| = 720/20 = 60$. Taking the same approach as in exercise 16, we arrive at the following list of inequivalent configurations.

(1, 2, 3, 4, 5, 6)	(1, 2, 3, 4, 6, 5)	(1, 2, 3, 5, 4, 6)	(1, 2, 3, 5, 6, 4)
(1, 2, 3, 6, 4, 5)	(1, 2, 3, 6, 5, 4)	(1, 2, 4, 3, 5, 6)	(1, 2, 4, 3, 6, 5)
(1, 2, 4, 5, 3, 6)	(1, 2, 4, 5, 6, 3)	(1, 2, 4, 6, 3, 5)	(1, 2, 4, 6, 5, 3)
(1, 2, 5, 3, 4, 6)	(1, 2, 5, 3, 6, 4)	(1, 2, 5, 4, 3, 6)	(1, 2, 5, 4, 6, 3)
(1, 2, 5, 6, 3, 4)	(1, 2, 5, 6, 4, 3)	(1, 2, 6, 3, 4, 5)	(1, 2, 6, 3, 5, 4)
(1, 2, 6, 4, 3, 5)	(1, 2, 6, 4, 5, 3)	(1, 2, 6, 5, 3, 4)	(1, 2, 6, 5, 4, 3)
(1, 3, 2, 4, 5, 6)	(1, 3, 2, 4, 6, 5)	(1, 3, 2, 5, 4, 6)	(1, 3, 2, 5, 6, 4)
(1, 3, 2, 6, 4, 5)	(1, 3, 2, 6, 5, 4)	(1, 3, 4, 2, 5, 6)	(1, 3, 4, 2, 6, 5)
(1, 3, 4, 5, 2, 6)	(1, 3, 4, 5, 6, 2)	(1, 3, 4, 6, 2, 5)	(1, 3, 4, 6, 5, 2)
(1, 3, 5, 2, 4, 6)	(1, 3, 5, 2, 6, 4)	(1, 3, 5, 4, 2, 6)	(1, 3, 5, 4, 6, 2)
(1, 3, 5, 6, 2, 4)	(1, 3, 5, 6, 4, 2)	(1, 3, 6, 2, 4, 5)	(1, 3, 6, 2, 5, 4)
(1, 3, 6, 4, 2, 5)	(1, 3, 6, 4, 5, 2)	(1, 3, 6, 5, 2, 4)	(1, 3, 6, 5, 4, 2)
(1, 4, 2, 3, 5, 6)	(1, 4, 2, 3, 6, 5)	(1, 4, 2, 5, 3, 6)	(1, 4, 2, 5, 6, 3)
(1, 4, 2, 6, 3, 5)	(1, 4, 2, 6, 5, 3)	(1, 4, 3, 2, 5, 6)	(1, 4, 3, 2, 6, 5)
(1, 4, 3, 5, 2, 6)	(1, 4, 3, 5, 6, 2)	(1, 4, 3, 6, 2, 5)	(1, 4, 3, 6, 5, 2)

19. There are $3^3 = 27$ 3-colorings of a triangle. The group of symmetries of an equilateral triangle is S_3 consisting of the identity, rotations through 120 and 240 degrees and the three reflections across altitudes. For the identity e all colorings are fixed, so $|X_e| = 27$. For a rotation, the vertices must all have the same color to remain fixed, hence $|X_r| = 3$ for each of the two rotations r. In a reflection, one vertex remains fixed and so can have any of threes color. The other two vertices must have the same color and so can be any of three colors. Therefore $|X_\tau| = 3 \cdot 3 = 9$ for each reflection τ. By Burnside's Lemma, the number R, of inequivalent colorings, satisfies $R|S_3| = 6R = 27 + 3 + 3 + 9 + 9 + 9 = 60$ so $R = 10$.

21. The group of symmetries of the square is D_4 consisting of the identity, rotations through 90, 180 and 270 degrees, two reflections across lines bisecting opposite sides and two reflections across diagonals. There $5^4 = 625$ possible 5-colorings of the vertices. Thus $|X_e| = 625$ for the identity e. The rotations through 90 and 270 degrees force all vertices to have the same color so $|X_r| = 5$ for those two rotations r while $|X_r| = 5^2 = 25$ for the rotation through 180 degrees since pairs of opposite vertices are mapped to each other. A reflection across a diagonal fixes the vertices the diagonal passes through so each may be colored in five ways while the pair of opposite vertices must have the same one of the five colors. Thus

$|X_\tau| = 5^3 = 125$ for each diagonal reflection τ. By Burnside's Lemma, the number R, of inequivalent colorings, satisfies $R|D_4| = 8R = 625 + 5 + 5 + 25 + 125 + 125 + 25 + 25 = 960$ so $R = 120$.

23. Picture a tetrahedron as a pyramid with a triangular base. For illustration purposes, label the vertex at the top of the pyramid 1 and the base vertices 2, 3 and 4. The group of symmetries is A_4 as described in the text. We are considering 3-colorings of which there are $3^4 = 81$ in all. the identity fixes all vertices so $|X_e| = 81$. There are two rotations through each axis of symmetry through a vertex and the center of the opposite face for a total of eight rotations. For example the two such rotations about the vertex 1 are (1)(234) and (1)(243). To fix a coloring, the vertex 1 may have any coloring while vertices 2, 3, 4 must be identically colored with any of the three colors. Thus $|X_r| = 3 \cdot 3 = 9$ for any of the eight rotations. The remaining four motions are rotations of 180 degrees about a line from the center of one edge to the center of an opposite edge. For example, the rotation corresponding to the line from the center of the edge 34 to the edge 12 is (12)(34). Thus 1 and 2 must have the same color and 3 and 4 must have the same color. Hence $|X_r| = 3 \cdot 3 = 9$ for these rotations as well. By Burnside's Lemma $R|A_4| = 12R = 81 + 8(9) + 3(9) = 180$ and so $R = 15$.

25. Here the group of symmetries is $G = S_3$. In Example 4 of Section 9.3 of the text, the cycle index of S_3 was computed to be $P_{S_3}(x1, x2, x3) = (1/6)(x_1^3 + 3x_1 x_2 + 2x_3)$. The number of inequivalent 3-colorings of the vertices of an equilateral triangle is $P_{S_3}(3,3,3) = (1/6)(3^3 + 3 \cdot 3 \cdot 3 + 2 \cdot 3) = 10$.

27. Here the group of symmetries is $G = D_5$. The elements of D_5 decomposed into cycles are (1)(2)(3)(4)(5), the rotations: (1,2,3,4,5), (1,3,5,2,4), (1,4,2,5,3), (1,5,4,3,2) and the reflections: (1)(2,5)(3,4), (2)(1,3)(4,5), (3)(2,4)(1,5), (4)(3,5)(1,2), (5)(1,4)(2,3). Using the notation of the definition of the cycle index, there is 1 element with $i_1 = 5$, 4 elements with $i_5 = 1$ and 5 elements with $i_1 = 1, i_2 = 2$. Since $|D_5| = 10$ the cycle index of G is $P_G(x_1, x_2, x_3, x_4, x_5) = (1/10)(x_1^5 + 4x_5 + 5x_1 x_2^2)$ and the number of inequivalent 2-colorings of the square is $P_G(2,2,2,2,2) = (1/10)(2^5 + 4 \cdot 2 + 5 \cdot 2 \cdot 2^2) = 8$.

29. The group of symmetries of the tetrahedron is $G = A_4$. The table below lists the elements of A_4 by type of motion factored into cycles. The corresponding term in the cycle index of G is also listed.

Element	Motion Type	Term in P_{A_4}
(1)(2)(3)(4)	identity	x_1^4
(1)(2,3,4), (1)(2,4,3), (2)(1,4,3), (2)(1,3,4), (3)(1,4,2), (3)(1,2,4), (4)(1,3,2), (4)(1,2,3)	rotation about line through a vertex and its opposite face	$8x_1 x_3$
(1,2)(3,4), (1,3)(2,4), (1,4)(2,3)	rotation about line through center of one edge and center of opposite edge	$3x_2^2$

Since A_4 has 12 elements, the cycle index of A_4 is given by $P_{A_4}(x_1, x_2, x_3, x_4) = (1/12)(x_1^4 + 8x_1 x_3 + 3x_2^2)$ and the number of inequivalent 3-colorings of the regular hexagon is $P_{D_6}(3,3,3,3) = (1/12)(3^4 + 8 \cdot 3 \cdot 3 + 3 \cdot 3^2) = 15$.

Section 9.3 Advanced Exercises

1. Let D denote the derangements of $\{1, 2, ... n\}$ and D_j, for $j = 1, 2, ... (n-1)$, those derangements in which n is mapped to j. Then $D_1, D_2, ... D_{n-1}$ partition the set D since n must map to one and only one of $1, 2, ... (n-1)$ in

98   Chapter 9 Combinatorics

any derangement. For a given j, Consider those elements in D_j for which j is mapped to n, that is, derangements which swap j and n. Such a derangement is determined exactly by specifying a derangement of the remaining $(n-2)$ elements $\{1,2,...,(j-1),(j+1),...(n-1)\}$. Thus there are $d(n-2)$ such derangements. Next consider those elements of that do not swap n and j, that is n maps to j but j does not map to n. Such a derangement is determined by a map $\{1,2,...(n-1)\} \mapsto \{1,2,...(j-1),(j+1)...n\}$ where no value maps to itself and j does not map to n. But this is equivalent to a derangement of $\{1,2,...(n-1)\}$ in which n plays the role of j. Hence there are $d(n-1)$ derangements in D_j that do not swap n and j. Hence $|D_j| = d(n-1) + d(n-2)$ and $d(n) = |D| = |D_1| + |D_2| + ... + |D_{n-1}| = (n-1)(d(n-1) + d(n-2))$.

3. Let S_n denote the set of permutations of the set $\{1,2,3,...,n\}$. Let P_k for $k = 0,1,2,...,n$ denote those permutations that fix exactly k elements. Then $P_0, P_1,..., P_n$ partition S_n. An element of P_k can be constructed by first selecting the k elements to be fixed, in $\binom{n}{k}$ ways, and then rearranging the remaining $n-k$ elements so there are no additional fixed points, that is, deranging the $n-k$ elements, which can be done in $d(n-k)$ ways. Thus $|P_k| = \binom{n}{k} d(n-k)$ and $n! = |S_n| = \sum_{k=0}^{n} |P_k| = \sum_{k=0}^{n} \binom{n}{k} d(n-k)$.

5. Recall $\frac{1}{e} = 1 - \frac{1}{1!} + \frac{1}{2!} - \frac{1}{3!} + ...$ Using the expression for $d(n)$ from Exercise 4, we have

$\left| d(n) - \frac{n!}{e} \right| = \left| (-1)^{n+1} \frac{1}{(n+1)!} + (-1)^{n+2} \frac{1}{(n+2)!} + ... \right| \leq \left| (-1)^{n+1} \frac{1}{(n+1)!} \right| = \frac{1}{(n+1)!} < \frac{1}{2}$. Since $d(n)$ is within $\frac{1}{2}$ of $\frac{n!}{e}$, $d(n)$ is the closest integer to $\frac{n!}{e}$.

7. First note $\sum_{k=0}^{n} k \binom{n}{k} d(n-k) = \sum_{k=1}^{n} k \binom{n}{k} d(n-k) = \sum_{k=1}^{n} k \frac{n!}{k!(n-k)!} d(n-k) = \sum_{k=1}^{n} \frac{n!}{(k-1)!(n-k)!} d(n-k)$

(by setting $j = k-1$) $= \sum_{j=0}^{n-1} \frac{n!}{j!(n-1-j)!} d(n-1-j) = n \sum_{j=0}^{n-1} \frac{(n-1)!}{j!(n-1-j)!} d(n-1-j) = n(n-1)! = n!$

where we used the fact the summation is the $(n-1)$ case of the formula in Exercise 3. If X represents the number of fixed points in a permutation in S_n, then by the counting argument used in the solution to Exercise 3, $P(X = k) = \dfrac{\binom{n}{k} d(n-k)}{n!}$ and so $E(X) = \sum_{k=0}^{n} k \cdot P(X = k) = \frac{1}{n!} \sum_{k=0}^{n} k \cdot \binom{n}{k} d(n-k) = \frac{1}{n!} \cdot n! = 1$.

Section 9.4 Supplementary Exercises

1. For the Stirling numbers of the first kind, the first five equations are

$x = x$
$x(x-1) = x^2 - x$
$x(x-1)(x-2) = x^3 - 3x^2 + 2x$

$$x(x-1)(x-2)(x-3) = x^4 - 6x^3 + 11x^2 - 6x$$
$$x(x-1)(x-2)(x-3)(x-4) = x^5 - 10x^4 + 35x^3 - 50x^2 + 24x$$

For the Stirling numbers of the second kind, the first five equations are

$$x = x$$
$$x^2 = x + x(x-1)$$
$$x^3 = x + 3x(x-1) + x(x-1)(x-2)$$
$$x^4 = x + 7x(x-1) + 6x(x-1)(x-2) + x(x-1)(x-2)(x-3)$$
$$x^5 = x + 15x(x-1) + 25x(x-1)(x-2) + 10x(x-1)(x-2)(x-3) + x(x-1)(x-2)(x-3)(x-4)$$

3. The triangle below gives the Stirling numbers of the second kind for $n=1$ to $n=8$ from top to bottom. Rows are arranged in order of increasing k, that is, a row consists of $S(n,1), S(n,2), ..., S(n,n)=1$.

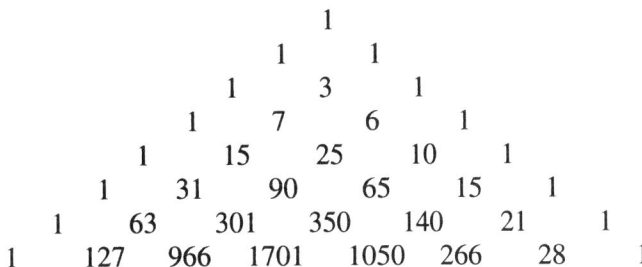

5. The table below lists the cycle decomposition for each permutation of the 5 elements in $\{1,2,3,4,5\}$. The first column lists those decompositions consisting of 1 cycle, the second 2 cycles, and so on. Note the number of permutations in each column are 24, 50, 35, 10 and 1 respectively corresponding to the values of $s(5,1), s(5,2), s(5,3), s(5,4),$ and $s(5,5)$.

(1,2,3,4,5)	(1)(2,3,4,5)	(1)(2)(3,4,5)	(1)(2)(3)(4,5)	(1)(2)(3)(4)(5)
(1,2,3,5,4)	(1)(2,3,5,4)	(1)(2)(3,5,4)	(1)(2)(3,4)(5)	
(1,2,4,5,3)	(1)(2,4,5,3)	(1)(2,3)(4,5)	(1)(2)(3,5)(4)	
(1,2,4,3,5)	(1)(2,4,3,5)	(1)(2,3,4)(5)	(1)(2,3)(4)(5)	
(1,2,5,4,3)	(1)(2,5,4,3)	(1)(2,3,5)(4)	(1)(2,4)(3)(5)	
(1,2,5,3,4)	(1)(2,5,3,4)	(1)(2,4,3)(5)	(1)(2,5)(3)(4)	
(1,3,4,5,2)	(1,2)(3,4,5)	(1)(2,4,5)(3)	(1,2)(3)(4)(5)	
(1,3,5,4,2)	(1,2)(3,5,4)	(1)(2,4)(3,5)	(1,3)(2)(4)(5)	
(1,3,2,4,5)	(1,2,3)(4,5)	(1)(2,5,3)(4)	(1,4)(2)(3)(5)	
(1,3,5,2,4)	(1,2,3,4)(5)	(1)(2,5,4)(3)	(1,5)(2)(3)(4)	
(1,3,2,5,4)	(1,2,3,5)(4)	(1)(2,5)(3,4)		
(1,3,4,2,5)	(1,2,4,3)(5)	(1,2)(3)(4,5)		
(1,4,5,3,2)	(1,2,4,5)(3)	(1,2)(3,4)(5)		
(1,4,3,5,2)	(1,2,4)(3,5)	(1,2)(3,5)(4)		
(1,4,5,2,3)	(1,2,5,3)(4)	(1,2,3)(4)(5)		
(1,4,2,3,5)	(1,2,5,4)(3)	(1,2,4)(3)(5)		
(1,4,2,5,3)	(1,2,5)(3,4)	(1,2,5)(3)(4)		
(1,4,3,2,5)	(1,3,2)(4,5)	(1,3,2)(4)(5)		
(1,5,4,3,2)	(1,3,4,2)(5)	(1,3)(2)(4,5)		
(1,5,3,4,2)	(1,3,5,2)(4)	(1,3,4)(2)(5)		
(1,5,4,2,3)	(1,3,4,5)(2)	(1,3,5)(2)(4)		

(1,5,2,3,4)	(1,3,5,4)(2)	(1,3)(2,4)(5)		
(1,5,2,4,3)	(1,3)(2,4,5)	(1,3)(2,5)(4)		
(1,5,3,2,4)	(1,3,2,4)(5)	(1,4,2)(3)(5)		
	(1,3,5)(2,4)	(1,4,5)(2)(3)		
	(1,3)(2,5,4)	(1,4)(2)(3,5)		
	(1,3,2,5)(4)	(1,4)(2,3)(5)		
	(1,3,4)(2,5)	(1,4)(2,5)(3)		
	(1,4,3,2)(5)	(1,5,2)(3)(4)		
	(1,4,5,2)(3)	(1,5,3)(2)(4)		
	(1,4,2)(3,5)	(1,5,4)(2)(3)		
	(1,4,5,3)(2)	(1,5)(2)(3,4)		
	(1,4,3,5)(2)	(1,5)(2,3)(4)		
	(1,4,2,3)(5)	(1,5)(2,4)(3)		
	(1,4,5)(2,3)	(1,4,3)(2)(5)		
	(1,4)(2,3,5)			
	(1,4,3)(2,5)			
	(1,4)(2,5,3)			
	(1,4,2,5)(3)			
	(1,5,3,2)(4)			
	(1,5,4,2)(3)			
	(1,5,2)(3,4)			
	(1,5,4,3)(2)			
	(1,5,3,4)(2)			
	(1,5,2,3)(4)			
	(1,5,4)(2,3)			
	(1,5)(2,3,4)			
	(1,5,3)(2,4)			
	(1,5)(2,4,3)			
	(1,5,2,4)(3)			

7. The table below lists the partitions of the set $\{1,2,3,4,5\}$. The first column lists those partitions consisting of 1 set, the second 2 sets, and so on. Note the number of partitions in each column are 1, 15, 25, 10 and 1 respectively corresponding to the values of $S(5,1)$, $S(5,2)$, $S(5,3)$, $S(5,4)$, and $S(5,5)$.

$\{1,2,3,4,5\}$	$\{1\}\{2,3,4,5\}$	$\{1\}\{2\}\{3,4,5\}$	$\{1\}\{2\}\{3\}\{4,5\}$	$\{1\}\{2\}\{3\}\{4\}\{5\}$
	$\{2\}\{1,3,4,5\}$	$\{1\}\{3\}\{2,4,5\}$	$\{1\}\{2\}\{4\}\{3,5\}$	
	$\{3\}\{1,2,4,5\}$	$\{1\}\{4\}\{2,3,5\}$	$\{1\}\{2\}\{5\}\{3,4\}$	
	$\{4\}\{1,2,3,5\}$	$\{1\}\{5\}\{2,3,4\}$	$\{1\}\{3\}\{4\}\{2,5\}$	
	$\{5\}\{1,2,3,4\}$	$\{2\}\{3\}\{1,4,5\}$	$\{1\}\{3\}\{5\}\{2,4\}$	
	$\{1,2\}\{3,4,5\}$	$\{2\}\{4\}\{1,3,5\}$	$\{1\}\{4\}\{5\}\{2,3\}$	
	$\{1,3\}\{2,4,5\}$	$\{2\}\{5\}\{1,3,4\}$	$\{2\}\{3\}\{4\}\{1,5\}$	
	$\{1,4\}\{2,3,5\}$	$\{3\}\{4\}\{1,2,5\}$	$\{2\}\{3\}\{5\}\{1,4\}$	
	$\{1,5\}\{2,3,4\}$	$\{3\}\{5\}\{1,2,4\}$	$\{2\}\{4\}\{5\}\{1,3\}$	
	$\{2,3\}\{1,4,5\}$	$\{4\}\{5\}\{1,2,3\}$	$\{3\}\{4\}\{5\}\{1,2\}$	
	$\{2,4\}\{1,3,5\}$	$\{1\}\{2,3\}\{4,5\}$		
	$\{2,5\}\{1,3,4\}$	$\{1\}\{2,4\}\{3,5\}$		
	$\{3,4\}\{1,2,5\}$	$\{1\}\{2,5\}\{3,4\}$		
	$\{3,5\}\{1,2,4\}$	$\{2\}\{1,3\}\{4,5\}$		
	$\{4,5\}\{1,2,3\}$	$\{2\}\{1,4\}\{3,5\}$		
		$\{2\}\{1,5\}\{3,4\}$		
		$\{3\}\{1,2\}\{4,5\}$		
		$\{3\}\{1,4\}\{2,5\}$		

{3}{1,5}{2,4}
{4}{1,2}{3,5}
{4}{1,3}{2,5}
{4}{1,5}{2,3}
{5}{1,2}{3,4}
{5}{1,3}{2,4}
{5}{1,4}{2,3}

9. There are 20 3-cycles (i,j,k) among the permutations of $\{1,2,3,4,5\}$ since there are $5 \cdot 4 \cdot 3 = 60$ sequences of the form (i,j,k) but $(i,j,k) = (j,k,i) = (k,i,j)$ are equal as permutations. Further all 3-cycles are even since $(i,j,k) = (i,k)(i,j)$ an even number of transpositions. Similarly there are 120 sequences (i,j,k,l,m) but for any fixed values of $i,j,k,l,$ and m, five of the sequences are equivalent as permutations $(i,j,k,l,m) = (j,k,l,m,i) = \ldots$ and so there 24 distinct 5-cycles all of which are even since $(i,j,k,l,m) = (i,m)(i,l)(i,k)(i,j)$. A product of disjoint transpositions $(i,j)(k,l)$ is always even. To see that there are 15 such distinct permutations, note that the pair (i,j) can be formed in $\binom{5}{2} = 10$ ways and the pair (k,l) then formed by selecting 2 elements from the remaining 3 in $\binom{3}{2} = 3$ ways. Thus there are $3 \cdot 10 = 30$ products $(i,j)(k,l)$ however we have counted $(i,j)(k,l)$ and $(k,l)(i,j)$ as distinct products but they are the same permutations. Thus there are 15 products of disjoint transpositions. Adding the identity e completes the classification of the elements of A_5.

11. The cycle index of A_5 follows immediately from the classification in Exercise 9 above. The identity gives rise to the term x_1^5, the 20 3-cycles to $20 x_1^2 x_3$, the 24 5-cycles to $24 x_5$ and the 15 products of disjoint transpositions yield $15 x_1 x_2^2$. Given $|A_5| = 5!/2 = 60$ the cycle index is then
$P_{A_5}(x_1, x_2, x_3, x_4, x_5) = (1/60)(x_1^5 + 20 x_1^2 x_3 + 24 x_5 + 15 x_1 x_2^2)$.

Section 9.4 Advanced Exercises

1. Reconstruction of the entire table is left to the reader. As an example, consider the permutation 1324. Start on the 1 and count "one". The 1 is removed immediately leaving 324. Start on the 3 and count "one, two". You are one the 2 so it is removed leaving 34. Start counting "one, two, three, four, one, two, three, four..." on the 4 and you will see you always call "three" when you are on the 4 and "four" when you are on the 3. Thus these two cards will never be removed. Hence the Cards Removed column contains the list 12.

Also note that by Advanced Exercise 4 in Section 9.3, the number of derangements of $\{1,2,3,4\}$ is $d(4) = 4!\left(1 - \frac{1}{1!} + \frac{1}{2!} - \frac{1}{3!} + \frac{1}{4!}\right) = 9$. These correspond to the nine tables entries where no cards are removed.

Chapter 10 Models of Computation

Section 10.1 Exercises

1. The set $(a+b)^*$ represents all words in a and b. The set $(a+b)^*ab$ must then consist of all words in a and b which end with the two character string ab.

3. The set $(00+11)^*$ consists of all binary strings formed by concatenating any number of 00 or 11 sequences together. Another way to describe $(00+11)^*$ is as the set of all binary strings containing only strings of 0's and 1s of even length such as 1100001100 and 0011111111.

5. The set of words $(a+b)^*(c)^*$ consists of any word in a and b followed by any number of c's such as *aabbaaccccc* and *bbbaabc*.

7. The set $(01+10)^*$ is the set of all words in 01 and 10, such as 01011001 and 100101100110.

9. The set $(a+bc)^*$ is the set of all words in a and bc such as *abc*, *bcbcaaaabc* and *aaabcabcbc*.

11. $L(G) = \{\varepsilon, ab, ab^2\}$ and $L(G)$ is regular.

13. The productions give $B = b, ba$ and $A = a, ab, aba$. Iterating $s ::= AsB$ gives $s ::= A^n sB^n$ for $n > 0$. The language $L(G)$ is regular and
$L(G) = \varepsilon + (a+ab+aba)(b+ba) + (a+ab+aba)^2(b+ba)^2 + (a+ab+aba)^3(b+ba)^3 + ...$

15. $L(G) = \{\varepsilon, a, b, ab\}$ and $L(G)$ is regular.

17. $L(G)$ is regular and $L(G) = \{\varepsilon, abc, accc, abbc, abccc, accbc, acccccc, bcbc, bcccc\}$.

19. The productions $B ::= a$ and $B ::= aB$ indicate B may be any sequence of a's. $A ::= bB$ and $A ::= b$ allows A to be a word beginning with b followed by a (possibly empty) sequence of a's. Lastly iterating $A ::= aA$ allows A of the form $L(G) = b(a)^*$ to be prefixed by any sequence of a's of length at least one. $L(G)$ is regular and $L(G) = \varepsilon + b + a(a)^* b(a)^*$.

Section 10.2 Exercises

1.

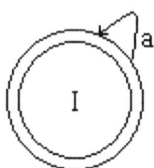

3.

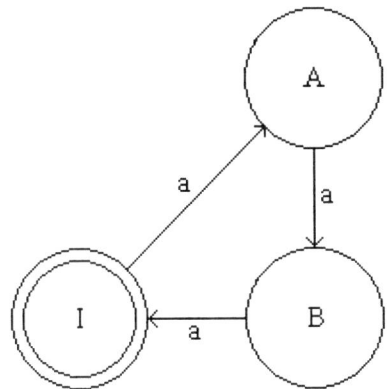

5.

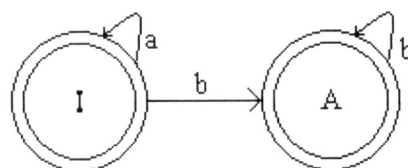

7.

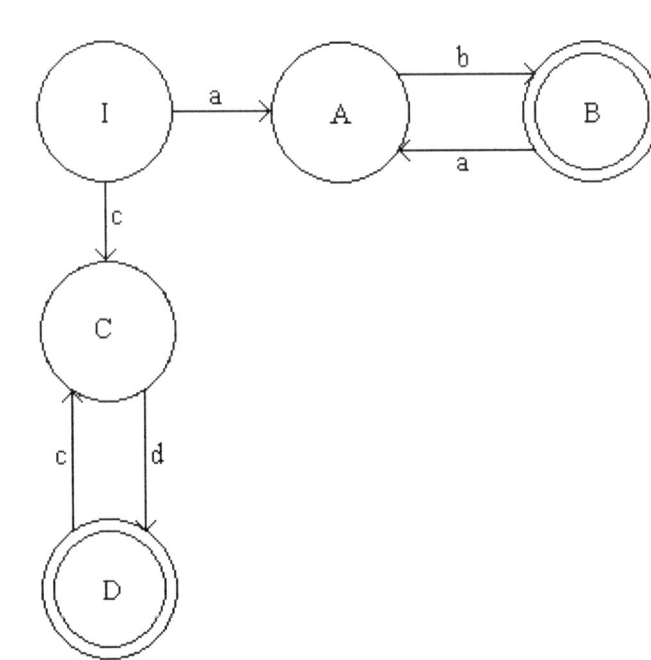

9.

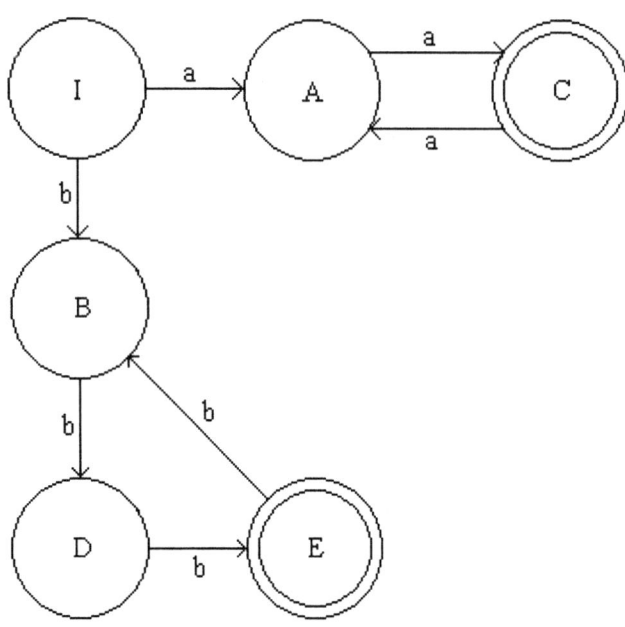

11. A path starting at I and ending at the accepting state B must begin with either a^2, ab or b. Once at the state B, the path may loop through any sequence of a's and b's. Thus the language M accepts is $(a^2 + ab + b)(a+b)^*$.

13. A path from I to I consists of repeated cycles of the form a^3. The language M accepts is then $(a^3)^*$.

15. A path from I to one of the accepting states C or D will start in either state A or B. If the path starts at A (a) it may then move indefinitely between B and A ending at A ($a(ba)^*$) before moving to state C ($a(ba)^* a = (ab)^* a^2$) or the path may move from A, alternate between A and B, then move to D from B ($a(ba)^* bb = (ab)^* b^2$). A similarly situation holds if B is the first state. Also note once in state C or D any combination of a's and b's may arise as the path moves back and forth from C to D or loops back into a state. The language M accepts is $((ba)^*(a+b^2) + (ab)^*(b+a^2))(a+b)^*$.

Section 10.3 Exercises

1. Suppose S is countable and $P \subseteq S$. The elements of S can be listed $s_1, s_2, s_3, \ldots$ and the elements of P must appear in this list in order given by the indices $i_1, i_2, \ldots$ say. Thus $s_{i_1}, s_{i_2}, s_{i_3}, \ldots$ is a list of the elements of P and P is countable.

3. Suppose $S_1, S_2, \ldots, S_r$ are countable sets. List the elements of S_i as $s_{i1}, s_{i2}, s_{i3}, \ldots$. The elements of $S_1 \times S_2 \times \ldots \times S_r$ can be listed as $(s_{11}, s_{21}, \ldots, s_{r1})$, $(s_{11}, s_{21}, \ldots, s_{r2})$, $(s_{11}, s_{21}, \ldots, s_{(r-1)2}, s_{r1}), \ldots$ and so on. Specifically, list all the ordered r-tuples where the second indices add to r. Then list those for which the second indices add to $r+1$, then $r+2$ and so on. For each value $n = r, r+1, r+2\ldots$ the pairs whose indices sum to n form a

finite set and can be listed using lex order on the subscripts Clearly this list contains all elements of $S_1 \times S_2 \times ... \times S_r$ and so $S_1 \times S_2 \times ... \times S_r$ is a countable set.

5. Suppose $S_1 \subset S_2 \subset ... S_k \subset ...$ is an ascending chain of countable sets. Thus each set S_k has a listing $s_{k1}, s_{k2}, s_{k3}, ...$ of its elements associated with it. Define a function $f : \bigcup_{i=1}^{\infty} S_i \to Z \times Z$ where Z is the set of integers as follows. Given $x \in \bigcup_{i=1}^{\infty} S_i$ then $x \in S_i$ for some i, let k be the smallest such value of i. Then x has a unique place on the list $s_{k1}, s_{k2}, s_{k3}, ...$, say $x = s_{km}$. Define $f(x) = (k, m)$. Then f is an injection since if $f(x) = f(y) = (k, m)$ then x and y are both equal to the mth element of S_k and so $x = y$. Let $R = f(\bigcup_{i=1}^{\infty} S_i) \subset Z \times Z$. Then f defines a one-one correspondence between $\bigcup_{i=1}^{\infty} S_i$ and R. This correspondence must have an inverse $g : R \to \bigcup_{i=1}^{\infty} S_i$ that is also a one-to-one correspondence. Now Z is a countable set and so, by the result of Exercise 2, is $Z \times Z$. By the result of Exercise 1, R is countable and lastly by Exercise 5, $\bigcup_{i=1}^{\infty} S_i$ is countable.

The solutions are written using Java and Maple. Java is a popular programming language; it was chosen for its expressiveness and portability. Maple is a well-known computer algebra system; it was used for problems requiring arbitrary-precision computations or symbolic algebraic expressions. Readers familiar with any high-level programming language should have no problems understanding the programs below.

Chapter 1

1.1.1 `for i from 0 to 40 do printf(`%d! = %d\n`,i,i!) od:`

1.1.3 `for i from 0 to 25 do printf(`%d -> %d\n`,i,binomial(2*i,i)) od:`

1.1.5

```
main := proc(n)  # Computes the n-th row of the Pascal triangle
    local a, prev, new, i, j;
    a := array(0 .. n + 1); a[0] := 1;
    for i from 1 to n do
        a[i] := 1; prev := 1;
        for j from 1 to i - 1 do new := a[j] + prev; prev := a[j]; a[j] := new od
    od;
    printf(`Row %d:\n`, n); for i from 0 to n do printf(`%d  `, a[i]) od; print();
end;
```

1.2.1 `s:=0: for i from 1 to 24 do s:=s+i*i; od: print(s,70*70):`

1.2.3

```
class PerfectSquares {
    public static void main(String[] args) {
        // Warning: for range > 1,650,000 arithmetic overflow will occur
        long range = 200000, y = 1;
        for (long x = 1; x <= range; x++) {
            long p = x*(x+1)*(2*x+1);
            while (6*y*y < p) y++;
            if (6*y*y == p)  System.out.println("x=" + x + "  y=" + y);
        }
    }
}
```

1.2.5

```
class TriangularNumbers {
    public static void main(String[] args) {
        // Warning: for range > 4,000,000 arithmetic overflow will occur
        long range = 200000, y = 1;
        for (long x = 1; x <= range; x++) {
```

```
                long p = x*(x+1)*(x+2);
                while (6*y*y < p) y++;
                if (6*y*y == p) System.out.println("x=" + x + "  y=" + y);
            }
        }
    }
}
```

1.3.1

```
class StandardGrayCode {
    public static void main(String[] args) {
        // Warning: for n>30 arithmetic overflow will occur
        int n = 10;                     // Vector length
        int m = (int) Math.pow(2,n);    // Number of vectors m = 2^n
        byte a[][] = new byte[m][n];    // Array to store all vectors
        a[0][0] = 0; a[1][0] = 1;
        int last = 2;
        for (int i = 1; i < n; i++) {
            // Compute vectors with length i+1. Vectors with length i are in rows 0 to last-1.
            for (int j = 0; j < last; j++) {
                // Replicate row j
                for (int k = 0; k < i; k++) a[2*last-1-j][k] = a[j][k];
                // For simplicity, put the extra bit at the end. The printing will be in reverse order.
                a[j][i] = 0; a[2*last-1-j][i] = 1;
            }
            last = 2*last;
        }
        for (int i = 0; i < m; i++) { System.out.println(); for (int j = n - 1; j >= 0; j--) System.out.print(a[i][j]); }
    }
}
```

1.3.5

```
class Knapsack {
    public static void main(String[] args) {
        // Warning: for n>30 arithmetic overflow will occur
        long N = 200;                   // Upper bound
        int n = 11;                     // Vector length
        int m = (int) Math.pow(2,n);    // Number of vectors m = 2^n
        byte a[][] = new byte[m][n];    // Array to store all vectors

        // First, generate all vectors of length n
        a[0][0] = 0; a[1][0] = 1;
        int last = 2;
        for (int i = 1; i < n; i++) {
            // Compute vectors with length i+1. Vectors with length i are in rows 0 to last-1.
            for (int j = 0; j < last; j++) {
                // Replicate row j
                for (int k = 0; k < i; k++) a[2*last-1-j][k] = a[j][k];
```

```
                // For simplicity, put the extra bit at the end
                a[j][i] = 0; a[2*last-1-j][i] = 1;
            }
            last = 2*last;
        }

        // Compute the solution to the knapsack problem
        byte current_sol[] = new byte[n];  // Current best solution
        long current_max = -1;              // Current best weight

        for (int i = 0; i < m; i++) {
            // The j-th element of the row has weight (j+1)*(j+2)/2, since the columns are
            // numbered [0,...,n-1] and the weights are [1,3,6,...]
            long weight = 0;
            for (int j = 0; j < n; j++) weight = weight + a[i][j]*(j+1)*(j+2)/2;
            if (weight <= N && weight > current_max) {
                current_max = weight; for (int j = 0; j < n; j++) current_sol[j] = a[i][j];
            }
        }

        System.out.print("Solution: ");
        for (int j = 0; j < n; j++) if (current_sol[j] == 1) System.out.print((j+1)*(j+2)/2 + " ");
        System.out.println("\nSum: " + current_max);
    }
}
```

1.4.1

```
class AllGrayCodes {
    public static void main(String[] args) {
        // The first row of "a" will be the first vector of the code, etc.
        int a[][] = new int[16][4];
        a[0][0] = a[0][1] = a[0][2] = a[0][3] = 0; // The first vector is (0,0,0,0)

        // The algorithm maintains a current partial code in rows 0,1,...,last (0 <= last <= 15) of array "a".
        // If last = 15, the current code is complete.
        int last = 0;

        // Each vector has 4 possible successors in the code. The 0th successor has the 0th bit inverted, etc.
        // For example, 0101 has 0th successor 1101, 1st successor 0001, 2nd successor 0111 and 3rd
        // successor 0100. The algorithm takes the last vector a[last] of the current partial code C and appends
        // the 0th successor of a[last] in order to obtain new code C'; then C' is extended in all possible ways.
        // After C' is completely processes, the 1st successor of a[last] is appended to C in order to obtain a
        // new code C", and C" is extended in all possible ways. The same process is repeated with the 2nd and
        // the 3rd successor of a[last].

        // Array "aux" is used to remember the last successor that was completely processed. Element aux[i] is
        // a number in the set {-1,0,1,2,3}. Its value gives the last successor of a[i] that was completely
        // processed. Value -1 means that no successor has been completely processed yet.
```

```java
        int aux[] = new int[16];
        for (int i = 0; i < 16; i++) aux[i] = -1;

        int num_codes = 0; // Total number of codes

        // If aux[0] == 3, then the last successor of (0,0,0,0) was completely processed and we are done
        while (aux[0] < 3) {

            if (last == 15) { // The vector is complete
                num_codes++; System.out.println("\nCode " + num_codes + ":");
                for (int k = 0; k <= last; k++)
                    System.out.println(a[k][0]+""+a[k][1]+""+a[k][2]+""+a[k][3]);
                last--;         // Return to the previous vector
                aux[last]++;    // Try next successor
                continue;
            }

            // Try to add a new vector after "last"
            if (aux[last] == 3) {// All successors have been processed
                last--;         // Return to the previous vector
                aux[last]++;    // Try next successor
                continue;
            }

            // Need to explore successor (aux[last]+1)
            a[last+1][0] = a[last][0]; a[last+1][1] = a[last][1];
            a[last+1][2] = a[last][2]; a[last+1][3] = a[last][3];

            a[last+1][aux[last]+1] = 1 - a[last+1][aux[last]+1]; // Invert bit (aux[last]+1)

            // If row a[last+1] is different from all other rows before it, we have to explore it.
            // Otherwise, try the next successor.
            boolean need_to_explore = true;
            for (int k = 0; k <= last; k++) { // Compare row a[last+1] with row a[k]
                if (a[k][0] == a[last+1][0] && a[k][1] == a[last+1][1] &&
                    a[k][2] == a[last+1][2] && a[k][3] == a[last+1][3]) need_to_explore = false;
            }
            if (need_to_explore) {
                last++;         // Append the successor
                aux[last] = -1; // Start with the 0th successor of the new last vector
            }
            else aux[last]++;   // Try next successor
        } // while (aux[0] < 3)
        System.out.println("\nTotal number of codes: " + num_codes);
    }
}
```

1.5.1

```
main := proc()
   local F, i;
   F := array(0 .. 119); F[0] := 0; F[1] := 1;  for i from 2 to 119 do F[i] := F[i - 1] + F[i - 2] od;
   for i from 0 to 59 do printf(`%d\t%d\n`, F[i], F[i + 60]) od
end;
```

1.5.3

```
main := proc()
   local n, x0, x1, x2, tmp;
   n := 3;  x0 := 0;  x1 := 1;  x2 := 1;
   while (x0 mod 10 <> 0 or x1 mod 10 <> 0 or x2 mod 10 <> 1) do
     tmp := x0 + x1 + x2; x0 := x1; x1 := x2; x2 := tmp;
     n := n + 1; # Here x2 contains the value of T[n]
   od;
   printf(`The period is %d\n`, n - 2)
end;
```

1.5.5

```
class Collatz {
    public static void main(String[] args) {
        int n = 1000;
        for (int i = 1; i <= n; i++) System.out.println(i + " --> " + seq_length(i));
        check_interval(65,67,27);
        check_interval(290,294,117);
        check_interval(840,845,41);
        check_interval(136696632,136696751,247);
    }
    // *********************************
    public static long seq_length(long initial_val) {
        long len = 0, current_val = initial_val;
        while (current_val != 1) {
            if (current_val % 2 == 0) current_val = current_val/2;
            else current_val = 3*current_val + 1;
            len++;
        }
        return len;
    }
    // ***********************************************
    public static void check_interval(long x, long y, long value) {
        // Checks if the sequences have length "value" for all initial values in the interval [x,y]
        for (long i = x; i <= y; i++)
            if (seq_length(i) != value) { System.out.println("\nError for i = " + i); System.exit(1); }
        System.out.println("Interval [" + x + "," + y + "] is OK");
    }
}
```

1.6.1

```
main := proc()
    local i;
    for i from 100 by 100 to 1000 do printf(`F[%d] = %d\n`, i, fib(i)) od
end;

fib := proc(n)
    local i, x0, x1, tmp;
    x0 := 0; x1 := 1; for i from 1 to n do tmp := x0 + x1; x0 := x1; x1 := tmp od;
    RETURN(x0)
end;
```

1.6.3

```
main := proc()
    local i;
    for i from 100 by 100 to 1000 do printf(`T[%d] = %d\n`, i, trib(i)) od
end;

trib := proc(n)
    local i, x0, x1, x2, tmp;
    x0 := 0; x1 := 0; x2 := 1;
    if (n < 2) then RETURN(0) fi;
    for i from 3 to n do tmp := x0 + x1 + x2; x0 := x1; x1 := x2; x2 := tmp od;
    RETURN(x2)
end;
```

1.6.5

```
main := proc(n)
    local i, j, C;
    C := array(0 .. n); C[0] := 1;
    for i from 1 to n do
        C[i] := 0; for j from 0 to i - 1 do C[i] := C[i] + C[j]*C[i - j - 1] od;
        printf(`C[%d] = %d\n`, i, C[i])
    od
end;
```

Chapter 2

2.1.1

```
main := proc(n)
    local i, m;
    # First, find the prime Mersenne numbers. This is not too expensive computationally.
```

```
    for i from 1 to n do
        m := 2^ithprime(i) - 1;  if isprime(m) then printf(`M[%d] = %d is prime\n`, ithprime(i), m) fi
    od;
    # Next, factor the composite Mersenne numbers. This is very expensive computationally.
    for i from 1 to n do
        m := 2^ithprime(i) - 1;
        if (not isprime(m)) then printf(`M[%d] = %d: `, ithprime(i), m); print(ifactor(m)); fi
    od
end;
```

2.1.3

```
main := proc(n)
    local s, m, i;
    m := 2^n - 1; s := 4;
    # The value of s grows extremely fast because of s := s*s - 2. However, we can compute it mod m.
    for i from 1 to n-2 do s := (s*s - 2) mod m od;
    if (s = 0) then printf(`M[%d] = %d is prime\n`, n, m);
    else printf(`M[%d] = %d is composite\n`, n, m);
    fi
end;
```

2.1.7

```
readlib(ifactors);  # Use library function "ifactors" for factorization
main := proc()
    local i, m, n, list, f1, f2, tmp, x;
    m := 1; for i from 1 to 19 do m := m*ithprime(i) od; m := m+1;
    list := ifactors(m); f1 := list[2][1][1]; f2 := list[2][2][1]; # First and second factors
    printf(`N[19] = %d = (%d)(%d)\n`, m, f1, f2);

    # The Fermat Factorization Algorithm finds integers x and y such that x*x - m = y*y. Since m = f1*f2, we
    # have x+y = max(f1,f2), x-y = min(f1,f2) and therefore x = (f1+f2)/2. The algorithm considers all integer
    # numbers between the square root of m and x. Each number corresponds to one step of the algorithm.
    n := ceil(evalf(sqrt(m),20)); x := (f1+f2)/2; printf(`Number of steps: %d\n`,x-n+1);

    # Assume a 500MHz machine and 10 clock cycles per step. In practice, one step is likely to take many more
    # clock cycles.
    printf(`Time in hours: %d\n`,floor(evalf((x-n+1)/(5*10^7*3600))));
end;
```

2.2.1

```
main := proc(n)
    local i, x0, x1, tmp, s;
    x0 := 0; x1 := 1; s := 0;

    for i from 1 to n do
        tmp := x0 + x1; x0 := x1; x1 := tmp; # Here x0 = F[n]
```

```
      s := s + x0;
   od;
   print(s,x0+x1-1); # Here x0+x1 = F[n]+F[n+1] = F[n+2]
   if (s <> x0+x1-1) then printf(`Error!!!\n`); fi;
end;
```

2.2.3

```
main := proc(n)
   local i, x0, x1, tmp;
   x0 := 0; x1 := 1; for i from 1 to n-1 do tmp := x0 + x1; x0 := x1; x1 := tmp; od;
   print(x0*(x0+x1) - x1*x1,(-1)^n);
   if (x0*(x0+x1) - x1*x1 <> (-1)^n) then printf(`Error!!!\n`); fi;
end;
```

2.2.5

```
main := proc(n)
   local i, x0, x1, y0, y1, tmp;
   x0 := 0; x1 := 1; # Fibonacci numbers
   y0 := 2; y1 := 1; # Lucas numbers
   for i from 1 to n-1 do
      tmp := x0 + x1; x0 := x1; x1 := tmp;
      tmp := y0 + y1; y0 := y1; y1 := tmp;
   od;
   print(y1,2*x0+x1); if (y1 <> 2*x0+x1) then printf(`Error!!!\n`); fi;
end;
```

2.2.7

```
main := proc(n)
   local i, x0, x1, tmp, s;
   x0 := 0; x1 := 1; for i from 1 to n-1 do tmp := x0 + x1; x0 := x1; x1 := tmp; od;
   s := (x1+x0)^2 - x0^2;
   for i from 1 to n do tmp := x0 + x1; x0 := x1; x1 := tmp; od;
   print(x1,s); if (x1 <> s) then printf(`Error!!!\n`); fi;
end;
```

2.2.9

```
main := proc(n,m)
   local i, x0, x1, tmp, s, fn0, fn1, fm0, fm1;
   x0 := 0; x1 := 1; s := 0;
   fn0 := 0; fn1 := 1; fm0 := 0; fm1 := 1;
   for i from 1 to m+n+1 do
      tmp := x0 + x1; x0 := x1; x1 := tmp;
      if (n = i) then fn0 := x0; fn1 := x1; fi;
      if (m = i) then fm0 := x0; fm1 := x1; fi;
   od;
```

```
    print(x0,fn0*fm0+fn1*fm1);
    if (x0 <> fn0*fm0+fn1*fm1) then printf(`Error!!!\n`); fi;
end;
```

2.2.11

```
main := proc(m,n)
  local v, i, last, x, pos, bound;
  if (m <= 0 or n <= 0) then RETURN(0); fi;

  # Array v contains the denominators, in sorted order. In the beginning, v = [n,n,...,n]. In Maple, array size
  # cannot be dynamically increased; thus, we use an array with a fixed large size. If the number of fractions is
  # bigger than 10000, an error will occur.
  v := array(1..10000); for i from 1 to m do v[i] := n; od;

  # Variable "bound" contains the index of the current last element of the array
  bound := m;

  # Variable "last" contains the index of last smallest repeated denominator. For example,
  # if v = [11,12,13,13,13,14,14], "last" will be 5. If there are no repeated denominators, "last" will be -1.
  if (m = 1) then last := -1; else last := m; fi;

  while (last <> -1) do
      x := v[last]; v[last] := x+1; # Replace x with x+1

      # Insert x*(x+1), while preserving the sorted order. The new denominator is at the end of the sorted array.
      bound := bound + 1; v[bound] := x*(x+1);

      # Update the value of last
      if (last > 2 and v[last-1] = x and v[last-2] = x) then last := last-1;
      else
          pos := last;
          while (pos < bound and v[pos] <> v[pos+1]) do pos := pos+1; od;
          if (pos = bound) then last := -1; # No repeated denominators were found
          else # Find the last occurrence of the repeated denominator
              while (pos < bound and v[pos] = v[pos+1]) do pos := pos+1; od;
              last := pos;
          fi;
      fi;
  od;
  printf(`%d/%d = 1/%d`, m, n, v[1]); for i from 2 to bound do printf(` + 1/%d`,v[i]); od;
  printf(`\n\nTotal number of fractions: %d\n`,bound);
end;
```

2.3.1

```
main := proc(n,m)  # Compute F[0] to F[n+2] mod m
  local i, x0, x1, tmp;
  x0 := 0; x1 := 1; printf(`F[0] mod %d = 0\n`, m);
```

```
   for i from 1 to n do
      tmp := (x0 + x1) mod m; x0 := x1; x1 := tmp; printf(`F[%d] mod %d = %d\n`, i, m, x0);
   od;
   printf(`F[%d] mod %d = %d\n`, n+1, m, x1);
   printf(`F[%d] mod %d = %d\n`, n+2, m, (x0+x1) mod m);
end;
```

2.3.5

```
main := proc(m)  # Compute T[i] mod m until the period is found
   local i, x0, x1, x2, tmp;
   x0 := 0; x1 := 1; x2 := 1; i := 1;
   while (x0 <> 0 or x1 <> 0 or x2 <> 1) do
      tmp := (x0 + x1 + x2) mod m; x0 := x1; x1 := x2; x2 := tmp; i := i+1; # Here x0 = T[i] mod m
   od;
   printf(`The period is %d\n`, i);
end;
```

2.3.7

```
main := proc(n)  # Check the first n+1 numbers
   local i, g;
   for i from 0 to n do
      g := 2^(2^i) + 1;
      if isprime(g) then
         printf(`2^2^%d + 1 = %d is prime\n`, i, g);
      else
         printf(`2^2^%d + 1 = %d is composite\n`, i, g); print(ifactor(g));
      fi;
   od;
end;
```

2.3.9

```
main := proc(P,Q,E,T)
    local R, A, D, V, S, U, i;
    R := (P-1)*(Q-1); V := P*Q;

    # Since GCD(E,R) = 1, there exist integers A and B such that A*E+B*R = 1. Procedure "euclid" returns A.
    # We want to find a positive D such that ED = 1 mod R. Since A can be negative, we cannot have D := A.
    # Furthermore, we want D to be as small as possible; therefore, we have D := A mod R.
    A := euclid(E,R); D := A mod R; printf(`D = %d\n`, D);

    printf(`Original message: %d\n`, T);
    # Use a special "inert operator" &^ for efficiency. See the Maple documentation for "mod".
    S := T &^ E mod V; printf(`Encrypted message: %d\n`, S);
    U := S &^ D mod V; printf(`Decrypted message: %d\n`, U);
end;
```

It can be shown that there exist integers A and B such that GCD(x,y) = A*x + B*y. Euclid's algorithm
constructs a sequence of integers a[0], a[1],... such that a[0] = x, a[1] = y and a[i] = a[i+1]*q[i+1] + a[i+2].
Consider the following sequences of integers: X[0] = 1, X[1] = 0,..., X[i+1] = X[i-1] – q[i]*X[i] and
Y[0] = 0, Y[1] = 1,..., Y[i+1] = Y[i-1] – q[i]*Y[i]. It can be shown that a[i] = X[i]*x + Y[i]*y.

```
euclid := proc(x,y)
    local a, b, c, q, X0, X1, tmpX, Y0, Y1, tmpY;
    a := x; b := y; X0 := 1; X1 := 0; Y0 := 0; Y1 := 1;
    while (b <> 0) do
        c := irem(a,b); # Remainder of a divided by b
        q := iquo(a,b); # Quotient of a divided by b
        a := b; b := c;
        tmpX := X0 - q*X1; X0 := X1; X1 := tmpX;
        tmpY := Y0 - q*Y1; Y0 := Y1; Y1 := tmpY;
    od;
    # Here GCD(x,y) = a = X0*x+Y0*y
    RETURN(X0);
end;
```

Chapter 3

3.2.1 print(2^(2^7));

3.3.1 n := 10; print(add(binomial(n,k)^2, k=0..n), binomial(2*n,n));

3.3.3 n := 10; r := 3; print(add(binomial(r+k,k), k=0..n), binomial(r+n+1,n));

3.3.5 n := 20; r := 3; print(add((-1)^k*binomial(n-r,k)*(n+1)/(r+k+1), k=0..n-r), 1/binomial(n,r));

Chapter 4

4.1.1

```
class CompleteGraph {
  public static void main(String[] args) {
    int n = 20;       // Number of nodes
    long num = 0;     // Number of edges
    for (int i = 1; i < n; i++) for (int j = i+1; j <= n; j++) { System.out.println("(" + i + "," + j + ")"); num++; }
    System.out.println("Total: " + num);
  }
}
```

4.1.5

```
class Octahedron {
  public static void main(String[] args) {
    int n = 20;        // Number of nodes
    long num = 0;      // Number of edges
    for (int i = 1; i < 2*n; i++)
      for (int j = i+1; j <= 2*n; j++) if ( j-i != n ) { System.out.println("(" + i + "," + j + ")"); num++; }
    System.out.println("Total: " + num);
  }
}
```

4.1.7

```
class Cube {
  public static void main(String[] args) {
    int n = 6;         // Dimension
    long num = 0;      // Number of edges
    long m = (long) Math.pow(2,n); // Number of nodes m = 2^n
    // The order of the nodes is as follows: each node has an index which is equal to the number whose binary
    // representation is the same as the binary vector labeling the node. For example, node 0 is the one with
    // label (0,0,..,0), and node (2^n-1) is the one with label (1,1,...,1).

    for (long i = 0; i < m-1; i++) {
      // Find nodes with index > i whose labels differ by exactly one bit. Need to find the zero bits in the binary
      // representation of i and flip them one by one. Let i = (bit[n-1],...,bit[1],bit[0]).
      long q = i, p = 1;
      for (int j = 0; j < n; j++) { // Here q = i/(2^j) and p = 2^j
        if (q % 2 == 0) { System.out.println("(" + i + "," + (i+p) + ")"); num++; // (i+p) is i with bit j flipped }
        q = q/2; p = p*2;
      }
    }
    System.out.println("Total: " + num);
  }
}
```

4.1.11

```
class RandomGraph {
  public static void main(String[] args) {
    int n = 100, m = 200;
    // For simplicity, use adjacency matrix representation, where adj[i][j] == true if and only if there is an edge
    // between nodes i and j. Since the nodes are numbered 1,..,n instead of the usual 0,..n-1, row 0 and column 0
    // are not used.
    boolean adj[][] = new boolean[n+1][n+1];
    // Clean up the matrix; not necessary in Java, but may be needed in other languages
    for (int i = 1; i <= n; i++) for (int j = 1; j <= n; j++) adj[i][j] = false;
```

```
    // Generate the random graph
    double sq2 = Math.sqrt(2), sq3 = Math.sqrt(3);
    for (int k = 1; k <= m; k++) {
      int C = (int) (((long) Math.floor(sq2*k*n/2)) % n);
      int D = (int) (((long) Math.floor(sq3*k*n/2)) % n);
      if (C != D) // Avoid loops
        adj[C+1][D+1] = adj[D+1][C+1] = true;
    }

    // Is it connected? To find out, start from node 1 and find the number of distinct nodes reachable from it.
    // Element reachable[i] will be true if and only if node i is reachable from node 1.
    boolean reachable[] = new boolean[n+1];
    int num_reachable = 1;
    reachable[1] = true; for (int i = 2; i <= n; i++) reachable[i] = false;

    // Keep the reachable nodes in a sequence, in the order they were discovered. Newly discovered nodes are
    // added at the end. This is just a simplistic implementation of a queue, which is used to perform breadth-first
    // traversal of the graph.
    int queue[] = new int[n];
    int head = 0, tail = 0;
    queue[0] = 1;

    while (head <= tail) {
      int current_node = queue[head]; head++;
      // Find all nodes adjacent to current_node and put the newly discovered at the end of the queue
      for (int k = 1; k <= n; k++)
        if (adj[current_node][k] && !reachable[k]) {
          reachable[k] = true; num_reachable++;
          tail++; queue[tail] = k;
        }
    }
    if (num_reachable == n) System.out.println("Connected graph");
    else System.out.println("Disconnected graph");
  }
}
```

4.2.1

```
class AllEulerianCycles {

  public static long total_num_cycles = 0;

  public static void main(String[] args) {
    // Generate the adjacency matrix for K5
    int n = 5;
    boolean[][] adj = new boolean[n][n];
    for (int i = 0; i < n; i++) for (int j = 0; j < n; j++) adj[i][j] = (i != j);
```

```
// The graph should be a connected Eulerian graph. Check that all nodes have non-zero even degrees.
for (int i = 0; i < n; i++) {
  int d = degree(adj,i); if (d == 0 || d % 2 == 1) { System.out.println("Invalid graph"); System.exit(1); }
}

// The graph has at most n*(n+1)/2 edges, which is an upper bound on the length of any Eulerian cycle
int trail[] = new int[1+n*(n+1)/2];

// Start with a specific edge. For example, pick one of the edges incident to node 0.
trail[0] = 0;
int succ;
for (succ = 1; succ < n; succ++) if (adj[0][succ]) { trail[1] = succ; break; }

// Remove the starting edge from the graph
remove_edge(adj,n,0,succ);

// Start building the trail
build_trail(adj,n,trail,1);
}
// *******************************************************************
public static void build_trail(boolean adj[][], int curr_num_nodes, int trail[], int trail_length) {
// Parameters:
// adj - adjacency matrix of the graph
// curr_num_nodes - number of non-isolated nodes in the graph
// trail - current trail, as a sequence of nodes
// trail_length - number of edges in the current trail

int curr_node = trail[trail_length]; // Last node

if ( isolated(adj,curr_node) ) { // Complete cycle
  total_num_cycles++; System.out.print("\nCycle " + total_num_cycles + ": ");
  for (int i = 0; i <= trail_length; i++)System.out.print(trail[i] + " ");
  return;
}

// Find all nodes that can be added to the trail, according to Fleury's algorithm
int all_succ[] = successors(adj,curr_num_nodes,curr_node);
// Explore all successors
for (int i = 0; i < all_succ.length; i++) {
  int succ = all_succ[i];
  trail[trail_length+1] = succ;
  int new_num_nodes = remove_edge(adj,curr_num_nodes,curr_node,succ);

  build_trail(adj,new_num_nodes,trail,trail_length+1);

  // Restore the edge for the next successor
  adj[curr_node][succ] = adj[succ][curr_node] = true;
  }
}
```

// ***
public static int[] successors(boolean adj[][], int curr_num_nodes, int node) {
  // Returns an array of possible successors of "node", where "node" is non-isolated. If there is at least one
  // non-bridge edge incident to "node", the other end nodes of all such non-bridge edges are stored in the
  // returned array. If all incident edges are bridge edges, their end nodes are stored in the return array.

  int n = adj.length, num_bridges = 0, num_non_bridges = 0;

  // If edge (node,i) is a bridge, element is_bridge_edge[i] will be true
  boolean[] is_bridge_edge = new boolean[n];
  for (int i = 0; i < n; i++) is_bridge_edge[i] = false;

  // Find all bridges
  for (int i = 0; i < n; i++) {
    if (! adj[node][i]) continue;
    if (is_bridge(adj,curr_num_nodes,node,i)) { num_bridges++; is_bridge_edge[i] = true; }
    else num_non_bridges++;
  }

  int[] res;
  if (num_non_bridges > 0) res = new int [num_non_bridges];
  else res = new int [num_bridges];
  int j = 0;
  for (int i = 0; i < n; i++) {
    if (! adj[node][i]) continue;
    if (num_non_bridges > 0) { if (! is_bridge_edge[i]) res[j++] = i; }
    else res[j++] = i; // Only bridge edges
  }
  return res;
}
// ***
public static int remove_edge(boolean adj[][], int curr_num_nodes, int x, int y) {
  // Removes edge (x,y). Returns the number of the remaining nodes. If the removal did not result in any new
  // isolated nodes, the return value is curr_num_nodes. Otherwise, it is the new number of remaining nodes.

  adj[x][y] = adj[y][x] = false;
  int ret_val = curr_num_nodes;
  if (isolated(adj,x)) ret_val--; // Node x becomes isolated
  if (isolated(adj,y)) ret_val--; // Node y becomes isolated
  return ret_val;
}
//***
public static boolean is_bridge(boolean adj[][], int curr_num_nodes, int x, int y) {
  // Variable curr_num_nodes contains the number of remaining nodes. Array adj is the adjacency matrix of the
  // graph. To determine if edge (x,y) is a bridge, remove it and check if the resulting graph is connected.

  adj[x][y] = adj[y][x] = false;
  boolean res = connected(adj,curr_num_nodes);
  adj[x][y] = adj[y][x] = true;

```
      return !res;
   }
   //*******************************************************
   public static boolean connected(boolean adj[][], int curr_num_nodes) {
      // Find out if the remaining nodes in the graph are connected. Variable curr_num_nodes contains the number
      // of remaining nodes. Array adj is the adjacency matrix of the graph. Starting from a non-isolated node, find
      // the number of nodes reachable from it. Return true if and only if this number is equal to curr_num_nodes.

      if (curr_num_nodes == 0) return true;

      int n = adj.length, start = 0;
      while (start < n && isolated(adj,start)) start++;
      if (start == n) return false;

      // Element reachable[i] is true if and only if node i is reachable from node "start"
      boolean reachable[] = new boolean[n];
      int num_reachable = 1;
      reachable[start] = true; for (int i = 0; i < n; i++) reachable[i] = false;

      // Keep the reachable nodes in a sequence, in the order they were discovered. Newly discovered nodes are
      // added at the end. This is just a simplistic implementation of a queue, which is used to perform breadth-first
      // traversal of the graph.
      int queue[] = new int[n];
      int head = 0, tail = 0;
      queue[0] = start;

      while (head <= tail) {
         int current_node = queue[head]; head++;
         // Find all nodes adjacent to current_node and put the newly discovered at the end of the queue
         for (int k = 0; k < n; k++)
            if (adj[current_node][k] && !reachable[k]) {
               reachable[k] = true; num_reachable++;
               tail++; queue[tail] = k;
            }
      }
      return (num_reachable == curr_num_nodes);
   }
   //*****************************************************
   public static boolean isolated(boolean adj[][], int node) {
      // Returns true if the node has no outgoing edges
      int n = adj.length; for (int i = 0; i < n; i++) if (adj[node][i]) return false;
      return true;
   }
   //***************************************************
   public static int degree(boolean adj[][], int node) {
      int res = 0; for (int i = 0; i < adj.length; i++) if (adj[node][i]) res++;
      return res;
   }
}
```

4.2.3 Use the same program as in 4.2.1, but with the following generation code:

```
// Generate the adjacency matrix for the 4-cube
int n = 16;
boolean adj[][] = new boolean[n][n];
for (int i = 0; i < n; i++) for (int j = 0; j < n; j++) adj[i][j] = false;
for (int i = 1; i <= 4; i++) adj[0][i] = true;
for (int i = 11; i <= 14; i++) adj[i][15] = true;
adj[1][5] = adj[1][6] = adj[1][8] = adj[2][5] = adj[2][7] = adj[2][9] = true;
adj[3][6] = adj[3][7] = adj[3][10] = adj[4][8] = adj[4][9] = adj[4][10] = true;
adj[5][11] = adj[5][12] = adj[6][11] = adj[6][13] = adj[7][11] = adj[7][14] = true;
adj[8][12] = adj[8][13] = adj[9][12] = adj[9][14] = adj[10][13] = adj[10][14] = true;
// Fill the lower part of the symmetric adjacency matrix
for (int i = 0; i < n; i++) for (int j = i+1; j < n; j++) adj[j][i] = adj[i][j];
```

4.2.5 Use the same program as in 4.2.1, but with the following generation code:

```
// Generate the adjacency matrix for K 2,2,2,2
int n = 8;
boolean[][] adj = new boolean[n][n];
for (int i = 0; i < n; i++) for (int j = 0; j < n; j++) adj[i][j] = (i != j);
adj[0][1] = adj[1][0] = adj[2][3] = adj[3][2] =
adj[4][5] = adj[5][4] = adj[6][7] = adj[7][6] = false;
```

4.2.7

```
class HamiltonianCircuits {
  public static void main(String[] args) {
    // The set of all Hamiltonian circuits in a complete graph can be obtained by generating all permutations of
    // the nodes. There are many algorithms for generating all permutations of n elements. The one presented
    // below uses a simplistic backtracking approach. For more sophisticated algorithms, see, for example,
    // "Combinatorial Algorithms" by Reingold, Nievergelt, and Deo, Prentice Hall, 1977.

    int n = 5;
    int a[] = new int[n]; // Array to store the permutations

    // The permutations will be generated by adding a valid node after the current partial permutation and
    // recursively trying to extend the resulting permutation. To keep track of the nodes that have already been
    // tried at position a[i], an extra array is used. Element aux[i] stores the last node that was at position a[i] and
    // for which all successor nodes have been completely explored. Value aux[i] == -1 shows that no such node
    // exists yet.
    int aux[] = new int[n]; for (int i = 0; i < n; i++) aux[i] = -1;

    // The current partial permutation is a[0],...,a[last]. The value of "last" is a number between 0 and n-1.
    // If it is n-1, this is a complete permutation.
    int last = 0;

    int num_circuits = 0; // Total number of circuits
    a[0] = 0;
```

```
  while (a[0] < n) {

    if (last == n-1) { // Complete permutation
      num_circuits++; System.out.print("\nCircuit " + num_circuits + ": ");
      for (int k = 0; k < n; k ++) System.out.print(a[k] + " ");
      last--;          // Return to the previous position
      aux[last]++;     // This successor was completely processed
      continue;
    }

    // Try to add a new node after "last"
    if (aux[last] == n-1) { // All successors have been processed
      if (last == 0) { a[0]++; aux[0] = -1; // New first element }
      else { last--; aux[last]++; }
      continue;
    }

    a[last+1] = aux[last] + 1; // Need to explore successor (aux[last]+1)
    // If a[last+1] is different from all nodes before it, we have to explore it. Otherwise, try the next successor.
    boolean need_to_explore = true;
    for (int k = 0; k <= last; k++) if (a[k] == a[last+1]) need_to_explore = false;
    if (need_to_explore) { last++; aux[last] = -1; }
    else aux[last]++;
  } // while (a[0] < n)
 }
}
```

4.3.1

```
with(linalg); # Include linear algebra package
main := proc()
     local a, n, deg, res;
     generate_graph('a','n');
     compute_degrees(a,n,'deg');
     modify_matrix(a,n,deg);
     # Function minor(a,1,1) returns the matrix obtained by removing the 1st row and the 1st column of a.
     # Function det computes the determinant of its argument.
     printf(`The number of spanning trees is %d\n`, det(minor(a,1,1)));
end;

# Generate the adjacency matrix for K 5,5
generate_graph := proc(adj,n)
     local i,j;
     n := 10; adj := matrix(10,10);
     for i from 1 to 10 do
        for j from 1 to 10 do if (i<=5 and j<=5 or i>5 and j>5) then adj[i,j]:=0; else adj[i,j]:=1; fi; od;
     od;
end;
```

124   SOLUTIONS TO COMPUTER EXERCISES

```
# Compute the degree of each node
compute_degrees := proc(adj,n,d)
      local i,j;
      d := array(1..n); for i from 1 to n do d[i] := 0;
      for j from 1 to n do if adj[i,j] = 1 then d[i]:=d[i]+1 fi; od;
      od;
end;

# Negate all elements and replace diagonal elements with node degrees
modify_matrix := proc(adj,n,deg)
      local i,j;
      for i from 1 to n do for j from 1 to n do if (i = j) then adj[i,j] := deg[i] else adj[i,j] := -adj[i,j] fi; od; od;
end;
```

4.3.3 Use the same program as in 4.3.1, but with the following generation procedure:

```
# Generate the adjacency matrix for the 8-wheel
generate_graph := proc(adj,n)
      local i,j;
      n := 9; adj := matrix(9,9); for i from 1 to 9 do for j from 1 to 9 do adj[i,j]:=0; od; od;
      for i from 2 to 9 do adj[i,1] := 1; adj[1,i] := 1; od;
      for i from 2 to 8 do adj[i,i+1] := 1; adj[i+1,i] := 1; od;
      adj[9,2] := 1; adj[2,9] := 1;
end;
```

4.3.5 Use the same program as in 4.3.1, but with the following generation procedure:

```
# Generate the adjacency matrix for the 12-node ladder graph
generate_graph := proc(adj,n)
      local i,j;
      n := 12; adj := matrix(12,12); for i from 1 to 12 do for j from 1 to 12 do adj[i,j]:=0; od; od;
      for i from 1 to 6  do adj[i,i+6] := 1; adj[i+6,i] := 1; od;
      for i from 1 to 11 do adj[i,i+1] := 1; adj[i+1,i] := 1; od;
      adj[6,7] := 0; adj[7,6] := 0;
end;
```

4.3.7 Use the same program as in 4.3.1, but with the following generation procedure:

```
# Generate the adjacency matrix for the icosahedron. The nodes are numbered according to the planar graph in
# Figure 4.16. Numbers increase from top to bottom. At the same height, numbers increase left to right.
generate_graph := proc(adj,n)
      local i,j;
      n := 12; adj := matrix(12,12); for i from 1 to 12 do for j from 1 to 12 do adj[i,j]:=0; od; od;
      for i from 2 to 4 do adj[1,i]:=1; od;
      for i from 4 to 6 do adj[3,i]:=1; od;
      adj[1,9]:=1;      adj[1,12]:=1;     adj[2,3]:=1;      adj[2,5]:=1;      adj[2,10]:=1;
      adj[2,12]:=1;     adj[4,6]:=1;      adj[4,7]:=1;      adj[4,9]:=1;      adj[5,6]:=1;
      adj[5,8]:=1;      adj[5,10]:=1;     adj[6,7]:=1;      adj[6,8]:=1;      adj[7,8]:=1;
      adj[7,9]:=1;      adj[7,11]:=1;     adj[8,10]:=1;     adj[8,11]:=1;     adj[9,11]:=1;
```

adj[9,12]:=1;   adj[10,11]:=1;   adj[10,12]:=1;   adj[11,12]:=1;
    for i from 1 to 11 do for j from i+1 to 12 do adj[j,i] := adj[i,j]; od; od; # Fill the lower part of the matrix
end;

4.3.9

Consider two planar graphs G_1 and G_2 such that: (1) there is a one-one correspondence between the set of nodes in G_1 and the set of faces in G_2 (counting the area outside of G_2 as an additional face); (2) there is an edge between two nodes in G_1 if and only if their corresponding faces in G_2 share an edge; (3) the above properties are also true in the opposite direction (i.e., the nodes in G_2 correspond to faces in G_1, etc.). Two such graphs are *dual*; it can be shown that in this case G_1 and G_2 have the same number of spanning trees. It is easy to see that the planar projections of the icosahedron and the dodecahedron (Figure 4.16) are dual.

4.3.11

```
class CubeSpanningTrees {
  public static void main(String[] args) {
    // Computes the set of all spanning trees of the 3-cube, represented by their Prufer codes
    int num_trees = 0; // Total number of trees
    boolean adj[][] = new boolean[8][8]; // Adjacency matrix for the 3-cube
    for (int i = 0; i < 8; i++) for (int j = 0; j < 8; j++) adj[i][j] = false;

    // Node with label (x,y,z) has index 4*x+2*y+z. There are 8 nodes and 12 edges.
    adj[0][1] = adj[0][2] = adj[0][4] = adj[1][3] = adj[1][5] = adj[2][3] =
    adj[2][6] = adj[3][7] = adj[4][5] = adj[4][6] = adj[5][7] = adj[6][7] = true;
    for (int i = 0; i < 8; i++) for (int j = i+1; j < 8; j++) adj[j][i] = adj[i][j];

    // Generate all 8^6 tuples of 6 numbers; each number is in the set {0,...,7}. Then use the Prufer
    // correspondence to get labeled trees and check if they are spanning trees for the 3-cube.
    int tuple[] = new int[6]; for (int i = 0; i < 6; i++) tuple[i] = 0;

    // The last position with value less than 7. If pos == -1, tuple == (7,7,7,7,7,7)
    int pos = 5;

    while (pos != -1) {
      // Check if the tree corresponding to the tuple is a spanning tree for the 3-cube
      if (is_spanning_tree(tuple,adj)) {
        num_trees++; System.out.print("\nTree " + num_trees + ": ");
        for (int i = 0; i < 6; i++) System.out.print((tuple[i]+1)); // Print the Prufer code of the tree
      }

      // Find the next tuple. First, find the last position with value less than 7.
      pos = 5; while (pos >= 0 && tuple[pos] == 7) pos--;
      if (pos == 1) continue;
      // Increment tuple[pos] and make all positions after it zero
      tuple[pos]++; for (int i = pos+1; i < 6; i++) tuple[i] = 0;
    }
  }
```

126   SOLUTIONS TO COMPUTER EXERCISES

// **
public static boolean is_spanning_tree(int tuple[], boolean adj[][]) {
  // Performs Prufer's mapping. Whenever an edge of the tree is found, checks if it is an edge in the 3-cube.

  boolean list[] = new boolean[8]; for (int i = 0; i < 8; i++) list[i] = true;
  int list_size = 8;

  // The algorithm removes elements from the beginning of the tuple. The remaining elements will be
  // tuple[6-tuple_size],...,tuple[5].
  int tuple_size = 6;

  while (list_size > 2) {
    // Find the smallest number in the list that is not in the tuple
    int smallest;
    for (smallest = 0; smallest < 8; smallest++)
      if (list[smallest] && ! in_tuple(tuple,tuple_size,smallest)) break;

    // There is an edge between smallest and tuple[6-tuple_size]. Is it an edge in the 3-cube?
    if (! adj[smallest][tuple[6-tuple_size]]) return false;

    list[smallest] = false; list_size--;   // Remove "smallest" from the list
    tuple_size--;   // Remove the first element of the tuple
  }

  // There is also an edge between the two remaining nodes in the list
  int first = -1, second = -1;
  for (int i = 0; i < 8; i++) if (list[i]) { if (first == -1) first = i; else second = i; }
  if (! adj[first][second]) return false;
  return true;
}
// **
public static boolean in_tuple(int tuple[], int size, int value) {
  for (int i = 6-size; i < 6; i++) if (tuple[i] == value) return true;
  return false;
}
}

Chapter 5

5.1.1 for n from 0 to 41 do if (not isprime(n*n-n+41)) then printf(`not prime for n = %d\n`,n); fi; od;

5.1.3

main := proc()
  local x, pi1, pi3;
  x := 2; pi1 := 0; pi3 := 0;

```
   while (1 = 1) do
      if (pi3 < pi1) then break fi;
      x := nextprime(x); if (x mod 4 = 1) then pi1 := pi1 + 1 else pi3 := pi3 + 1 fi
   od;
   printf(`The smallest value is %d\n`, x); printf(`pi1 = %d and pi3 = %d\n`, pi1, pi3);
end;
```

5.3.1

```
// This program uses a Vector data structure from the standard Java library
import java.util.Vector;
class Parenthesize {
  public static void main(String args[]) {
    int n = 7;                         // Number of letters
    char letter[] = new char[n+1];     // letter[0] is 'a', letter[1] is 'b', etc.
    char c = 'a';
    for (int i = 0; i < n; i++, c++) letter[i] = c;

    // Compute a vector of strings, each containing one possible way to parenthesize n letters
    Vector result = parenthesize(0,n-1,letter);
    for (int i = 0; i < result.size(); i++)
       System.out.println((i+1) + ": " + result.elementAt(i));
    System.out.println("\nCatalan number: " + n_choose_k(2*(n-1),n-1)/n);
  }
  // *********************************************
  public static Vector parenthesize(int i, int j,char letter[]) {
    // Returns a vector with all possible ways to parenthesize the numbers in the interval [i,j], where 0
    // corresponds to "a", 1 corresponds to "b", etc.

    Vector res = new Vector();                   // Returned vector
    StringBuffer str = new StringBuffer();       // Scratch space

    if (i == j)   { str.append(letter[i]); res.addElement(str.toString()); return res; }
    if (i == j-1) { str.append(letter[i]); str.append(letter[j]); res.addElement(str.toString()); return res; }

    // The interval contains more than two numbers
    Vector v1,v2;

    for (int k = i; k < j; k++) {

      // Separately process intervals [i,k] and [k+1,j]
      v1 = parenthesize(i,k,letter);
      v2 = parenthesize(k+1,j,letter);

      // Merge the results
      for (int p = 0; p < v1.size(); p++)
        for (int q = 0; q < v2.size(); q++) {
          String s1 = (String) v1.elementAt(p);
          String s2 = (String) v2.elementAt(q);
```

```
            if (s1.length() > 1) { str.append("("); str.append(s1); str.append(")"); }
            else str.append(s1);
            if (s2.length() > 1) { str.append("("); str.append(s2); str.append(")"); }
            else str.append(s2);
            res.addElement(str.toString()); str.setLength(0);
         }
      }
      return res;
   }
   // *******************************
   public static long n_choose_k(int n, int k) {
      long res;
      int m; if (k < n-k) m = k; else m = n-k;
      res = 1; for (int i = 1; i <= m; i++) res = (res*(n-m+i))/i;
      return res;
   }
}
```

Chapter 6

6.1.1

$x+y+z = x(y+\overline{y}) + y(z+\overline{z}) + z(x+\overline{x}) = xy(z+\overline{z}) + x\overline{y}(z+\overline{z}) + yz(x+\overline{x}) + y\overline{z}(x+\overline{x}) + xz(y+\overline{y}) + \overline{x}z(y+\overline{y}) =$
$xyz + xy\overline{z} + x\overline{y}z + x\overline{y}\,\overline{z} + \overline{x}yz + \overline{x}y\overline{z} + \overline{x}\,\overline{y}z$

6.1.3 The problem can be solved similarly to 6.1.1. Alternatively, we can consider all tuples (x,y,z,w) for which the expression is equal to 1. Each such (x,y,z,w) tuple corresponds to one minterm in the disjunctive normal form. The only tuple for which the value is 0 is $(0,0,0,0)$; therefore, every minterm except $\overline{x}\,\overline{y}\,\overline{z}\,\overline{w}$ is in the disjunctive normal form.

6.1.5 Similarly to 6.1.3, the answer is the sum of all minterms except $\overline{u}\,\overline{v}\,\overline{x}\,\overline{y}\,\overline{z}\,\overline{w}$.

6.1.7

$xyz + x\overline{y}z + xy\overline{z} + x\overline{y}\,\overline{z} + \overline{x}yz + \overline{x}y\overline{z} = xy(z+\overline{z}) + x\overline{y}(z+\overline{z}) + \overline{x}y(z+\overline{z}) = xy + x\overline{y} + \overline{x}y + \overline{x}y = x(y+\overline{y}) + y(x+\overline{x}) = x+y$

Chapter 8

8.1.1

```
class MergeSort {
   public static void main(String[] args) {
```

```java
    int[] a = { 21, 12, 14, 5,  47, 50, 33, 17, 1,  94, 51, 77, 85, 15, 19, 23, 59, 16, 93, 20,
                55, 18, 12, 35, 28, 66, 24, 90, 80, 70, 44, 56, 34, 27, 30, 31, 18, 67, 73,  58 };

    int[] res = mergesort(a,0,a.length-1);

    System.out.println("\nOriginal Array: "); for (int j = 0; j < a.length; j++)   System.out.print(a[j] + " ");
    System.out.println("\nSorted Array: ");   for (int j = 0; j < res.length; j++) System.out.print(res[j] + " ");
  }
  //************************************
  public static int[] mergesort(int[] a, int p, int q) {
    // Returns a[p]...a[q] in sorted order. The result is returned in a new array "ret".
    int[] ret = new int[q-p+1];
    if (p == q) { ret[0] = a[p]; return ret; } // Nothing to sort

    // Sort the two halves recursively
    int[] seq1 = mergesort(a,p,(p+q)/2);
    int[] seq2 = mergesort(a,(p+q)/2+1,q);

    // Merge
    int i = 0, j = 0, k = 0;
    while (j < seq1.length && k < seq2.length) {
      if (seq1[j] <= seq2[k])    { ret[i] = seq1[j]; i++; j++; }
      else                       { ret[i] = seq2[k]; i++; k++; }
    }

    // Take care of the remaining elements. Only one of these loops will execute.
    while (j < seq1.length) { ret[i] = seq1[j]; i++; j++; }
    while (k < seq2.length) { ret[i] = seq2[k]; i++; k++; }
    return ret;
  }
} // class MergeSort

//************
class QuickSort {
  public static void main(String[] args) {
    int[] a = { 21, 12, 14, 5,  47, 50, 33, 17, 1,  94, 51, 77, 85, 15, 19, 23, 59, 16, 93, 20,
                55, 18, 12, 35, 28, 66, 24, 90, 80, 70, 44, 56, 34, 27, 30, 31, 18, 67, 73,   58 };
    System.out.println("\nOriginal Array: "); for (int j = 0; j < a.length; j++)  System.out.print(a[j] + " ");
    quicksort(a,0,a.length-1);
    System.out.println("\nSorted Array: ");   for (int j = 0; j < a.length; j++)  System.out.print(a[j] + " ");
  }
  //************************************
  public static void quicksort(int[] a, int p, int q) {
    // Sorts a[p]...a[q]

    if (p >= q) return; // Nothing to sort
```

```
    // Partition
    int x = a[p], i = p-1, j = q+1, tmp;
    while (i < j) {
      do j--; while (a[j] > x);
      do i++; while (a[i] < x);
      if (i < j) { tmp=a[i]; a[i]=a[j]; a[j]=tmp; }
    }
    // Recursively sort the two parts
    quicksort(a,p,j);
    quicksort(a,j+1,q);
  }
}
```

8.1.3

```
class BinarySearch {
  public static void main(String[] args) {
    int[] a = {  1,  5,  12, 12, 14, 15, 16, 17, 18, 18, 19, 20, 21, 23, 24, 27, 28, 30, 31, 33,
                34, 35, 44, 47, 50, 51, 55, 56, 58, 59, 66, 67, 70, 73, 77, 80, 85, 90, 93,  94 };

    if (bin_search(a,35))  System.out.println("\nFound\n");
    else                   System.out.println("\nNot found\n");
  }
  //***********************************
  public static boolean bin_search(int[] a, int x) {
    int i = 0, j = a.length-1;
    while (i <= j) {
      int k = (i+j)/2;
      if (x < a[k]) j = k-1;
      else if (x > a[k]) i = k+1;
      else return true; // Element found
    }
    return false; // Element not found
  }
}
```

8.1.7

```
class RevLexOrder {
  public static void main(String[] args) {
    // Prints all k-element subsets of a set. Each subset is considered as a sorted k-tuple of distinct elements of the
    // original set. The subsets are printed in revlex order. For simplicity, the original set is represented using a
    // sorted array.

    int n = 5;                                  // Set size
    int a[] = new int[n];                       // The set
    for (int i = 0; i < n; i++) a[i] = i+1;     // Running example
    int k = 3;                                  // Subset size
```

// Element b[i] contains the index of the element of "a" that is in position i in the subset
int b[] = new int[k];

// First subset in revlex order: (a[0],a[1],...,a[k-1])
for (int i = 0; i < k; i++) b[i] = i;

int total = 0; // Total number of subsets

while (b[0] < n-k) {
  total++; System.out.print("\nSubset " + total + ": {");
  for (int j = 0; j < k; j ++) System.out.print(a[b[j]] + ",");
  System.out.print("\b}");

  // Construct the next subset. First, find the smallest i such that b[i] can be incremented. Element b[i] can be
  // incremented only if b[i+1] - b[i] > 1.
  int i;
  for (i = 0; i < k-1; i++) if (b[i+1] > b[i] + 1) break;
  b[i]++;

  // Next, make b[0],...,b[i-1] as small as possible
  for (int j = 0; j < i; j++) b[j] = j;
}

// Print the last subset
total++; System.out.print("\nSubset " + total + ": {");
for (int j = 0; j < k; j ++) System.out.print(a[b[j]] + ",");
System.out.print("\b}");
 }
}

8.1.9

class GrevLexOrder {
  public static void main(String[] args) {
    // Prints all k-element subsets of a set. Each subset is considered as a sorted k-tuple of distinct elements of the
    // original set. The subsets are printed in grevlex order. For simplicity, the original set is represented using a
    // sorted array.

    int n = 10;   // Set size
    int a[] = new int[n];   // The set
    for (int i = 0; i < n; i++) a[i] = i+1; // Running example
    int k = 4;   // Subset size

    // ****************************** PHASE 1 ******************************
    // The algorithm first generates all k-subsets in revlex order and stores them in array "revlex"

    int m = n_choose_k(n,k);   // Number of subsets
    int revlex[][] = new int[m][k]; // All k-subset in revlex order

// For the subset stored as revlex[i], sum[i] contains the sum of its elements
int sum[] = new int[m];

// Array "b" is used during the construction of array "revlex". Element b[i] contains the index of the element
// of "a" that is in position i in the subset.
int b[] = new int[k];

// First subset in revlex order: (a[0],a[1],...,a[k-1])
for (int i = 0; i < k; i++) b[i] = i;

int total = 0; // Total number of currently generated subsets

while (b[0] < n-k) {
 // Store the current subset in array "revlex"
 sum[total] = 0;
 for (int j = 0; j < k; j++) { revlex[total][j] = b[j]; sum[total] = sum[total] + a[b[j]]; }
 total++;

 // Construct the next subset. First, find the smallest i such that b[i] can be incremented. Element b[i] can be
 // incremented only if b[i+1] - b[i] > 1.
 int i;
 for (i = 0; i < k-1; i++) if (b[i+1] > b[i] + 1) break;
 b[i]++;

 // Next, make b[0],...,b[i-1] as small as possible
 for (int j = 0; j < i; j++) b[j] = j;
}

// Remember the last subset
sum[total] = 0;
for (int j = 0; j < k; j++) { revlex[total][j] = b[j]; sum[total] = sum[total] + a[b[j]]; }

// *********************************** PHASE 2 ***************************************
// Next, the algorithm constructs an array "res" such that revlex[res[i]] is the i-th subset in grevlex order
int res[] = new int[m];

// First, find the number of subsets for each possible sum of elements. The smallest sum is sum[0], and the
// biggest sum is sum[m-1]. Array num[0,...,sum[m-1]-sum[0]] contains this information. Element num[i]
// contains the number of subsets with sum of elements equal to i+sum[0].
int range = sum[m-1] - sum[0] + 1;
int num[] = new int[range];
for (int i = 0; i < range; i++) num[i] = 0;
for (int i = 0; i < m; i++) num[sum[i]-sum[0]]++;

// All subsets with the same sum of elements form a subsequence. For each possible sum, we can compute the
// starting position of its subsequence. This information is stored in array "start".
int start[] = new int[range];
start[0] = 0; for (int i = 1; i < range; i++) start[i] = start[i-1] + num[i-1];

// Finally, construct an array "res" such that revlex[res[i]] is the i-th subset in grevlex order. Go through the
// subsets in revlex order, and store each one in the subsequence corresponding to its sum of elements.
// Element aux[i] contains the current number of subsets in the subsequence starting at start[i].
int aux[] = new int[range];
for (int i = 0; i < range; i++) aux[i] = 0;

for (int i = 0; i < m; i++) {
  int x = sum[i] - sum[0];
  res[start[x] + aux[x]] = i;
  aux[x]++;
}

for (int i = 0; i < m; i++) {
  System.out.print("\nSubset " + (i+1) + ": {");
  for (int j = 0; j < k; j ++) System.out.print(a[revlex[res[i]][j]] + ",");
  System.out.print("\b}");
}
}
// *******************************
public static int n_choose_k(int n, int k) {
  int res, m;
  if (k < n-k) m = k; else m = n-k;
  res = 1; for (int i = 1; i <= m; i++) res = (res*(n-m+i))/i;
  return res;
}
}

8.3.1 n:=1: while(1000*n^2+7000*n+10000 > n^3) do n:=n+1 od: print(n):

8.3.5 n:=1: while(evalf(1000*n*log[2](n)+20000*n) > n^2) do n:=n+1 od: print(n):

8.3.7 n:=1: while(n^50+n^40+n^30+n^20+n^10+1 > 2^n) do n:=n+1 od: print(n):

8.3.9

Let the input sequence be $S = a_0, a_1, \ldots, a_{n-1}$. Consider a subsequence $T = a_k, \ldots, a_{k+l}$ that appears both in S and the reverse of S. There exists a subsequence $T' = a_{m-l}, \ldots, a_m$ such that $a_m = a_k, a_{m-1} = a_{k+1}, \ldots, a_{m-l} = a_{k+l}$. The *center* of the pair (T, T') can be defined as follows: If $(k+m)$ is even, the center is the symbol with index $(k+m)/2$; if $(k+m)$ is odd, the center is between the symbols with indices $(k+m-1)/2$ and $(k+m+1)/2$. For example, for S = "abxyzbaz", T = "ab", and T' = "ba", the center is symbol "y".

class MaxCommon {
  public static void main(String[] args) {
    String input = "abxyzbazyxb";
    char[] a = input.toCharArray(); // Use a character array for simplicity
    int n = a.length;
    // The current longest common sequence is a[best_left]...a[best_right]
    int best_left = 0, best_right = 0;

```
// Find the longest (S1,S2) centered around each symbol.
for (int c = 0; c < n; c++) { // The center is a[c]
  int m; if (c < n-c-1) m = c; else m = n-c-1; // m = min(c,n-c-1)

  while (m > 0) {
    // Find a value of m such that a[c-m] == a[c+m]
    while (a[c-m] != a[c+m]) m--;
    if (m <= 0) break;

    // Find the subsequence starting at a[c-m] that also starts symmetrically at a[c+m]
    int k = 0;
    while (c-m+k < n && c+m-k >= 0 && a[c-m+k] == a[c+m-k]) k++;
    k--;

    // Is it longer that the current best subsequence?
    if (k > best_right - best_left) { best_left = c-m; best_right = c-m+k; }

    m = m-k-1; // Try to find the next subsequence
  }
}

// Next, consider each center that is between symbols
for (int c = 1; c < n; c++) { // The center is between a[c-1] and a[c]
  int m; if (c-1 < n-c-1) m = c-1; else m = n-c-1; // m = min(c-1,n-c-1)

  while (m >= 0) { // Find a value of m such that a[c-1-m] == a[c+m]
    while (m >= 0 && a[c-1-m] != a[c+m]) m--;
    if (m < 0) break;

    // Find the subsequence starting at a[c-1-m] that also starts symmetrically at a[c+m]
    int k = 0;
    while (c-1-m+k < n && c+m-k >= 0 && a[c-1-m+k] == a[c+m-k]) k++;
    k--;

    // Is it longer that the current best subsequence?
    if (k > best_right - best_left) { best_left = c-1-m; best_right = c-1-m+k; }

    m = m-k-1; // Try to find the next subsequence
  }
}
System.out.println("Input string: " + input); System.out.print("Longest common subsequence: ");
if (n > 0) for (int i = best_left; i <= best_right; i++) System.out.print(a[i]);
}
}
```

Complexity analysis: During each iteration of the first c-loop, $\min(c,n-c-1)$ pairs of symbols have to be compared. Therefore, the total number of operations for the first c-loop is proportional to $1 + 2 + \ldots + (n-1)/2 + n/2 + n/2 + \ldots + 1$ which is $\Theta(n^2)$. Similarly, the total number of operations for the second c-loop is $\Theta(n^2)$. Therefore, the overall complexity of the algorithm is $\Theta(n^2)$; this is both a lower and an upper bound.

8.4.1

```
main := proc(n) # Computes the sum of the first n terms
      local i, sum, bound, prec;
      prec := 20; # Number digits for computer arithmetic precision
      sum := 0; bound := evalf(Pi^2/6,prec);
      for i from 1 to n do sum := evalf(sum + 1/i^2,prec); od;
      printf(`Absolute error:\n\n`); print(evalf(bound-sum,prec));
      printf(`Relative error:\n\n`); print(evalf((bound-sum)/bound,prec));
end;
```

Chapter 9

9.1.1

```
main := proc(n)
      local f, k, d;
      f := tan(x); printf(`tan(x) = `);
      for k from 0 to n do
            d := eval(subs(x=0,f)); # The value of the k-th derivative at x = 0
            if (d <> 0) then printf(`%d*x^%d/%d! + `, d, k, k); fi;
            f := diff(f,x); # The derivative of f with respect to x
      od;
      printf(`...\n`);
end;
```

9.1.3

```
main := proc(n)
      local curr, prev, i, j;
      curr := array(0..n); prev := array(0..n); # Current and previous rows
      prev[0] := 1; printf(`A[0] = 1\n`);
      for i from 1 to n do
            if (i mod 2 = 1) then # Go left to right
                  curr[0] := 0; for j from 1 to i do curr[j] := curr[j-1] + prev[j-1]; od;
                  printf(`A[%d] = %d\n`, i, curr[i]);
            else # Go right to left
                  curr[i] := 0; for j from i-1 to 0 by -1 do curr[j] := curr[j+1] + prev[j]; od;
                  printf(`A[%d] = %d\n`, i, curr[0]);
            fi;
            for j from 0 to i do prev[j] := curr[j] od;
      od;
end;
```

136 SOLUTIONS TO COMPUTER EXERCISES

9.2.1.

```
# For each m from 0 to n, use the binomial inversion formula to compute P(B(m) - B(m+1))
main := proc()
     local n, m, k, sum;
     n := 10;
     for m from 0 to n do
          sum := 0;
          for k from m to n do
               sum := sum + (-1)^(k-m)*binomial(n,k)*binomial(k,m)*(1/(k+1));
          od;
          print(m,sum);
     od;
end;
```

9.3.1

```
main := proc(n) # Compute d(1),...,d(n)
     local i, x0, x1, tmp;
     x0 := 0; x1 := 1; printf(`d(1) = 0\nd(2) = 1\n`);
     for i from 3 to n do tmp := (i-1)*(x0 + x1); x0 := x1; x1 := tmp; printf(`d(%d) = %d\n`, i, x1); od;
end;
```

9.3.3

```
main := proc(m)
     local T, n, k, d, x;
     T := array(0..m); T[0] := 0; T[1] := 1;
     printf(`T[0] = 0\nT[1] = 1\n`);
     for n from 1 to m-1 do # Compute T[n+1]
          T[n+1] := 0;
          for k from 1 to n do
               x := 0; for d from 1 to k do if (k mod d = 0) then x := x + d*T[d]; fi; od;
               T[n+1] := T[n+1] + x*T[n-k+1];
          od;
          T[n+1] := T[n+1]/n; printf(`T[%d] = %d\n`, n+1, T[n+1]);
     od;
end;
```

9.4.1

If there are $n-1$ cards already removed, only one card is being compared against one, two, etc. Eventually, the card will agree with the count and will be removed.

Chapter 10

10.2.1

We will define a Turing machine that takes as input a string on the alphabet $\{0,1\}$. In the beginning, the read-write head is positioned on the leftmost symbol of the input string. If the string is a palindrome, the machine halts with a blank tape. If the string is not a palindrome, the machine halts with a non-blank tape, and the head is positioned on a non-blank symbol.

The tape alphabet S consists of 0, 1, and a special blank symbol denoted by b. The set of states Q contains a start state q_0, a halt state q_H, and several other states described below. The behavior of the machine can be completely specified by the transition function $t: (Q-\{q_H\}) \times S \rightarrow Q \times S \times \{R, L\}$. For current state q_i, current symbol s_j, and $t(q_i, s_j) = (q_k, s_m, R)$, the machine replaces s_j with s_m and changes the state to q_k. If the new state is q_H, the machine halts; otherwise, the head moves one position to the right (if the last element of the triple is L, the head moves to the left instead).

Below we present a description of the transfer function for all states reachable from initial configurations $(q_0, 0)$ and (q_0, b). It is trivial to complete the definition for initial configuration $(q_0, 1)$.

$t(q_0, b) = (q_H, b, R)$; palindrome
$t(q_0, 0) = (q_1, 0, R)$

$t(q_1, 0) = (q_1, 0, R)$
$t(q_1, 1) = (q_1, 1, R)$
$t(q_1, b) = (q_2, b, L)$

$t(q_2, 0) = (q_3, b, L)$; delete the last symbol
$t(q_2, 1) = (q_H, 1, L)$; not a palindrome
$t(q_2, b) = (q_H, b, R)$; this transition never occurs

$t(q_3, 0) = (q_3, 0, L)$
$t(q_3, 1) = (q_3, 1, L)$
$t(q_3, b) = (q_4, b, R)$

$t(q_4, 0) = (q_0, b, R)$; delete the first symbol
$t(q_4, b) = (q_H, b, R)$; palindrome
$t(q_4, 1) = (q_H, b, R)$; this transition never occurs

10.3.1

```
# Should be invoked with a string enclosed in backquotes. For example, main(`(~p)`); returns the Godel number
# of the sentence (~p). For simplicity, single symbols are used to encode AND, OR and -> (see below).
main := proc(s)
    local godel, i, res;
    # Use a table to store the Godel numbers
    godel[`|`] := 2;    # Use "|" for "OR"
    godel[`&`] := 3;    # Use "&" for "AND"
    godel[`-`] := 4;    # Use "-" for "->"
    godel[`~`] := 1;    godel[`(`] := 5;    godel[`)`] := 6;
    godel[`p`] := 7;    godel[`q`] := 11;   godel[`r`] := 13;

    res := 1;
```

```
        for i from 1 to length(s) do
            res := res * ithprime(i)^godel[substring(s,i)];
            printf(`(%d^%d)`, ithprime(i), godel[substring(s,i)]);
        od;
        printf(` = %d\n`, res);
end;
```

10.3.11

```
readlib(ifactors);
main := proc(n)
    local inv_godel, i, list, res;
    # Use an array to store the inverse mapping from Godel numbers to symbols
    inv_godel = array(1..13);
    # Numbers without a corresponding symbol are represented by ` ?`
    for i from 1 to 13 do inv_godel[i] := `?` od;
    inv_godel[1] := `~`;        inv_godel[2] := ` OR `;     inv_godel[3] := ` AND `;
    inv_godel[4] := ` -> `;     inv_godel[5] := `(`;        inv_godel[6] := `)`;
    inv_godel[7] := `p`;        inv_godel[11] := `q`;       inv_godel[13] := `r`;

    # "ifactors" returns a list in the form [1,[[p1,q1],[p2,q2],...,[pn,qn]]], where list[2][i][1] is the i-th prime in
    # the factorization and list[2][i][2] is the corresponding degree.
    list := ifactors(n);

    res := ``;
    for i from 1 to nops(list[2]) do # For each factor
        if (list[2][i][1] <> ithprime(i) or  list[2][i][2] > 13 or inv_godel[list[2][i][2]] = `?`) then
            printf(`\nINCORRECT NUMBER\n`); RETURN(0);
        fi;
        res := cat(res,inv_godel[list[2][i][2]]);
    od;
    print(res);
end;
```